OXFORD CHEMISTRY MASTERS

OXFORD CHEMISTRY MASTERS

1 A. Rodger and B. Nordén: *Circular dichroism and linear dichroism*
2 N. K. Terrett: *Combinatorial chemistry*
3 H. M. I. Osborn and T.H. Khan: *Oligosaccharides: Their chemistry and biological roles*
4 H. Cartwright: *Intelligent data analysis in science*
5 S. Mann: *Biomineralization: Principles and concepts in bioinorganic materials chemistry*

Biomineralization
Principles and Concepts in Bioinorganic Materials Chemistry

STEPHEN MANN

School of Chemistry, University of Bristol, Bristol UK.

This book has been printed digitally and produced in a standard specification in order to ensure its continuing availability

OXFORD
UNIVERSITY PRESS

Great Clarendon Street, Oxford OX2 6DP

Oxford University Press is a department of the University of Oxford.
It furthers the University's objective of excellence in research, scholarship,
and education by publishing worldwide in

Oxford New York

Auckland Cape Town Dar es Salaam Hong Kong Karachi
Kuala Lumpur Madrid Melbourne Mexico City Nairobi
New Delhi Shanghai Taipei Toronto
With offices in
Argentina Austria Brazil Chile Czech Republic France Greece
Guatemala Hungary Italy Japan South Korea Poland Portugal
Singapore Switzerland Thailand Turkey Ukraine Vietnam

Published in the United States
by Oxford University Press Inc., New York

Reprinted 2005

ISBN 0-19-850882-4

Antony Rowe Ltd., Eastbourne

Preface

The study of chemistry at the interface with other disciplines, such as biology, physics, materials science, environmental science or engineering, continues to be a source of much scientific creativity. Two recent success stories in this endeavour are *bioinorganic chemistry* and *materials chemistry*. Both areas are now represented by large international research communities, and routinely taught to final-year undergraduate and postgraduate chemistry students in many universities throughout the world. This book, which addresses the principles and concepts of *biomineralization*, lies at the interface between bioinorganic and materials chemistry, and therefore embraces and extends current notions of both these disciplines. Because the chemical aspects of biomineralization are generally absent or only briefly discussed in current textbooks, it seems timely to provide an overview of the field that is suitable for students and research scientists encountering the subject for the first time.

For this reason, I have tried to write a book that discusses the formation, structure and properties of biominerals, such as bones, shells, teeth, iron storage proteins, diatom shells and coccolith scales, principally from a chemical perspective, focusing on the overarching principles and concepts. The book is therefore organized on a thematic rather than a descriptive basis, so that the concepts and principles remain centre-stage throughout. This means that, except for Chapter 2, there is no attempt to survey the field on a subject-by-subject criterion. By focusing on the fundamental ideas and principles rather than addressing a shopping list of subject areas, I hope that the reader will be rewarded with an integrated understanding of biomineralization, and that the subject area can be covered in a coherent fashion within a short lecture course. To facilitate this further, I have included summary statements at the end of each section of the book, as well as a section at the end of each chapter that provides an overview containing the key points discussed. Of course, one immediate disadvantage of my approach is that the student will not find, for example, a comprehensive discussion of bone mineralization in one single section of the book because various aspects of the same subject area are used to illustrate different general principles or concepts. I have therefore cross-referenced the related sections to alleviate this fragmentation, and included a section on further reading at the end of each chapter.

The book begins with a short introductory chapter followed in Chapter 2 by a brief survey of the major types of biominerals. The general principles of biomineralization are discussed in Chapter 3 to set the scene for the next five chapters that describe the fundamental processes of biomineralization. These include the chemical control of biomineralization (Chapter 4), boundary-organized biomineralization (Chapter 5), organic matrix-mediated biomineralization (Chapter 6), morphogenesis (Chapter 7) and biomineral tectonics (Chapter 8). The chapters are best read in sequence as they are cumulative, especially the chapter on biomineral tectonics that assimilates ideas developed in Chapters 5 to 7. The last chapter of the book discusses biomineral-inspired approaches in materials chemistry and attempts to illustrate how current knowledge of biomineralization, as described in the preceding chapters, is inspiring new strategies in the synthesis of biomimetic materials. My apologies to other researchers in the field for using numerous examples in Chapter 9 that arise predominantly from my own research activities—this is not meant to infer that they represent the lion's share! The selection was made principally on the basis of how they dovetail with the themes established in the earlier chapters of the book.

I am indebted to many colleagues for their support and help in writing this book. I thank Professors R. B. Frankel, P. M. Harrison, N. H. Mendelson and S. Weiner, and Drs B. S. C. Leadbeater, C. C. Perry and J. R. Young, as well as my research group past and present, for the use of photographs and drawings arising from their research work. The book would not have been finished on time without the help of Mei Li who drew most of the figures. I am also indebted to Professor R. J. P. Williams FRS, who gave me the opportunity and inspiration to study biomineralization over 20 years ago at the University of Oxford. Finally, as I quickly discovered, book writing for a practising scientist is often squeezed into evenings and weekends, and this would not have been possible without the love and support of my wife Cindy and my children Jake and Rosa.

Stephen Mann
Bristol
August 2001

For Cindy, Jake, and Rosa

Contents

1 Inorganic structures of life

From the nano-world of rusty proteins and magnetic compasses in bacteria to the macroscopic structures of oyster shells, corals, ivory, bone and enamel, biology has evolved a new type of chemistry that brings together the synthesis and construction of hard and soft matter for the design of functionalized inorganic–organic materials. The process that gives rise to these small and large inorganic-based structures of life is called *biomineralization.*

Biomineralization: the study of the formation, structure and properties of inorganic solids deposited in biological systems.

1.1 Biomineralization—the big picture

Biomineralization involves the selective extraction and uptake of elements from the local environment and their incorporation into functional structures under strict biological control (Fig. 1.1). This process holds a special position in the science of life because unlike other biological tranformations, which leave no lasting signature on the environment or a tenuous scribble at best, the formation of hard bioinorganic materials such as bones and shells is unequivocally recorded in the fossil record. Both life and the environment have been fundamentally changed by the advent of biomineralization, which got going under full steam about 570 million years ago when skeletal hard parts became *de rigueur*. There is evidence—in the form of large concretions called *stromatolites*—that biological processes were involved in inorganic mineralization stretching as far back as 3500 million years (the earth is approximately 4200 million years old). These structures, however, appear to be the result of adventitious chemical combinations rather than the controlled deposition of inorganic solids for some precise biological function.

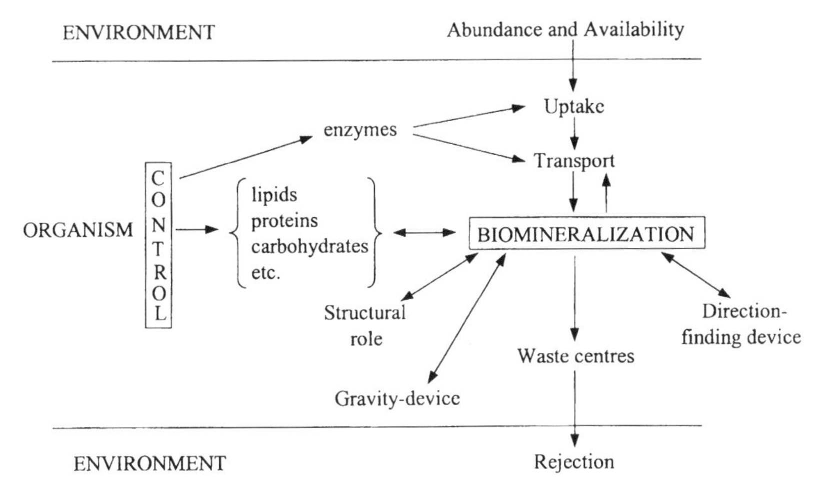

Fig. 1.1 Biomineralization and the environment.

The sudden proliferation in the number and types of shells and micro-skeletons made of minerals such as calcium carbonate, calcium phosphate or silica over half a billion years ago has had far-reaching implications on the global scale. Biomineralization is therefore an important aspect of many areas of the *earth sciences* (see further reading), such as:

- the global cycling of elements
- sedimentology
- fossilization (palaeontology and taxonomy)
- marine chemistry
- geochemistry.

For example, elements such as calcium, iron, carbon, phosphorus and silicon are cycled over millions of years through complex pathways that at some critical stage involve biomineralization (Fig. 1.2). Indeed, biomineral production is big business on the geological time scale, giving rise, for example, to huge chalk deposits such as the white cliffs of Dover on the south coast of Britain. These sedimentary rocks are the result of calcification processes in minute single-celled organisms that lived some 200 million years ago in a relatively warm shallow sea. Many of these biomineral structures remain as intact fossils even after compaction, such as shown in Fig. 1.3. Moreover, the fossils contain a record not only of the distant biology but also of the local climate and chemical conditions of the marine environment now long gone. Geochemists can extract this information in various ways. For example, paleotemperatures can be estimated by measurement of the $^{18}O/^{16}O$ isotopic ratio of fossilized shells. And the amounts of trace metals such as strontium incarcerated into the calcium carbonate shells of molluscs provide information about the salinity of the ancient seawater.

The evolution of biomineralization has provided organisms with a strong and tough building material. Whereas a tough skeleton can be made solely

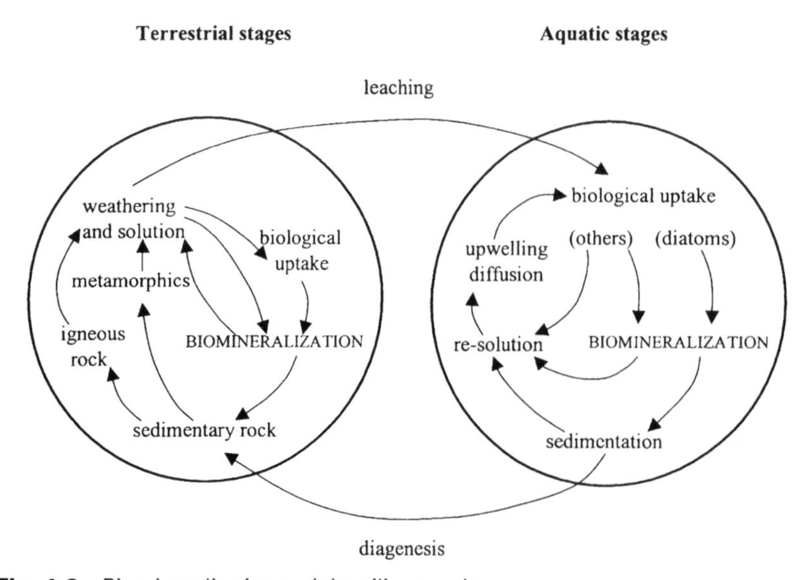

Fig. 1.2 Biomineralization and the silicon cycle.

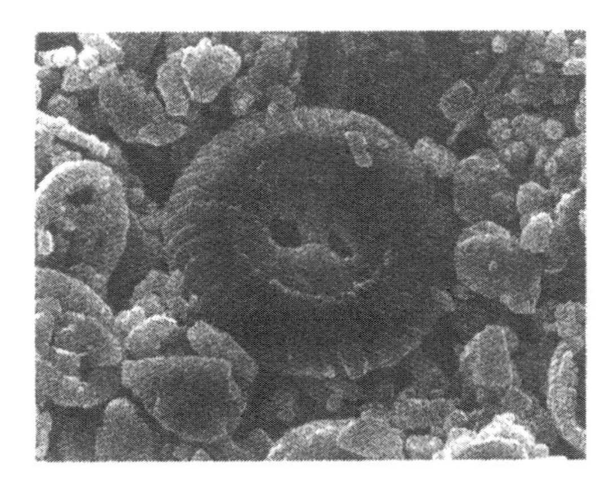

Fig. 1.3 Ancient biomineral structures (coccoliths) preserved as chalk.

from an organic biopolymer—the insect cuticle, for example, consists of a polysaccharide called *α-chitin*—the energy demand is high. Moreover, organic armour resists a bashing or hammering but doesn't stand up well when squeezed in the arms of a predator. Compared with an inorganic mineral, which is hard and stiff but brittle, organic materials are relatively soft and pliable but tough. So there is much to be gained in the mechanical design of life if the organic toughness is married with inorganic strength. A reasonable solution is to build a lightweight organic frame to save on metabolic energy and fill it with a cheap inorganic material such as calcium carbonate to produce an *inorganic–organic hybrid material*, or *biocomposite*, with well-defined mechanical properties.

But biomineralization offers an organism more than just structural support and mechanical strength. As Nature's master builder, it is involved in a wide variety of important biological functions such as:

- protection
- motion
- cutting and grinding
- buoyancy
- optical, magnetic and gravity sensing
- storage.

The types of minerals associated with these functions are described in more detail in the next chapter. From a wider perspective, we note that these higher-order functions arise from the evolution of specialized *tissues*, and that these structures must be integrated into the body as a whole if they are to function efficiently. Among these, the internal skeleton surely ranks as the most amazing example of an integrated biomineralized machine (Fig. 1.4). Clearly, the fundamental importance of this hard tissue means that there are serious medical consequences when the biomineralization process goes awry. Although of great importance, the study of pathological mineralization (kidney and urinary stones, dental calculus, etc.), as well as crippling diseases such as osteoporosis, is outside the scope of this book.

In summary, the big picture of biomineralization is one that contains many different subjects and perspectives, ranging from the global aspects of the

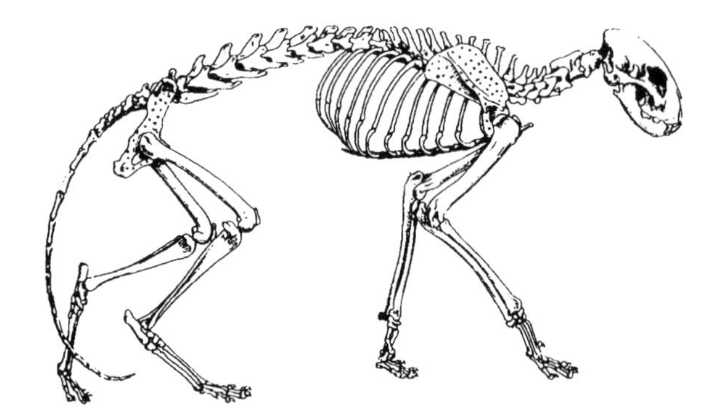

Fig. 1.4 The vertebrate skeleton.

earth sciences to the local niches of biology and the selection pressures on materials design, and to the anatomy of tissues and the microscopic world of cells. This diminishing length scale brings us ultimately to the molecular level and with it a requirement to understand the chemistry that drives biomineralization.

The processes of biomineralization have important consequences for evolution and the environment. Their impact is recorded on the global scale and stretches far back in the history of life. The study of biomineralization is therefore an interdisciplinary discipline, principally at the interface between chemistry, biology and materials science, but with important spin-offs for the fields of palaeontology, marine chemistry, sedimentology, medicine and dentistry.

1.2 Biomineralization—a new chemistry

Over the past two decades, the focus of biomineralization studies has shifted towards a chemical perspective, first as a new branch of *bioinorganic chemistry* and more recently as an important subject in the emerging field of *biomimetic materials chemistry*. For bioinorganic chemists, biomineralization represents an extension in the length scale of the interplay between biological processes and inorganic chemistry. Whereas biocoordination chemistry principally focuses on interactions between metal atoms and ligands at the level of the coordination sphere, biomineralization addresses the chemistry between collections of inorganic atoms (nucleation clusters, crystal faces, etc.) and multiple arrays of ligands arranged across organic surfaces (insoluble proteins, lipid membranes, etc.). The research aims of biomineralization in the context of bioinorganic chemistry include:

- the structural and compositional characterization of biominerals
- understanding the functional properties of biominerals
- elucidation of the processes through which organic macromolecules and organic structures control the synthesis, construction and organization of inorganic mineral-based materials.

At a more generic level, biomineralization is of key importance—along with other fields such as supramolecular chemistry—in the development of a modern paradigm of chemistry based on the concept that molecular-based interactions can be integrated into higher levels of organization and dynamics. This notion of *organized-matter chemistry* is particularly significant for biomineralization and the related field of *biomimetic materials chemistry* because they are each involved with the science of chemical construction of higher-order structures with increased levels of complexity.

The inorganic-based structures of life—biominerals—represent a new area of study for bioinorganic chemistry and are a source of inspiration in materials chemistry. Biomineralization is an example of organized-matter chemistry, which is concerned with the chemical construction, synthesis and emergence of organized architectures and complex forms.

1.3 This book

The main focus of this book is on the principles and concepts that arise from a chemical perspective of biomineralization. After surveying the major types of biominerals (Chapter 2), we discuss the general principles of biomineralization (Chapter 3), and then move on to describing the chemical aspects of biomineralization (Chapter 4). The next four chapters are concerned with the processes of biomineralization, including boundary-organized biomineralization (Chapter 5), organic matrix-mediated biomineralization (Chapter 6), morphogenesis (Chapter 7) and biomineral tectonics (Chapter 8). Chapter 9 describes how current knowledge of biomineralization is inspiring new biomimetic strategies in synthetic materials chemistry.

This book describes the principles and concepts of biomineralization and their application in the new field of biomimetic materials chemistry.

Further reading

Lowenstam, H. A. and Weiner, S. (1989). *On biomineralization*, pp. 207–251. Oxford University Press, New York.

Reynolds, C. S. (1986). Diatoms and the geochemical cycling of silicon. In *Biomineralization in lower plants and animals* (ed. Leadbeater, B. S. C. and Riding, R.), pp. 269–289. Systematics Association Vol. 30, Oxford University Press, Oxford.

Simkiss, K. and Wilbur, K. (1989). *Biomineralization: cell biology and mineral deposition*, pp. 299–313. Academic Press, San Diego.

2 Biomineral types and functions

Of the 20 to 25 essential elements required by living organisms, H, C, O, Mg, Si, P, S, Ca, Mn and Fe are common constituents of over 60 different biological minerals. Among these, calcium has a special place since it is not only exceedingly widespread but also the common constituent of familiar skeletal structures such as bones and shells. It is interesting to note at the outset that whereas bones are composed of calcium phosphate, shells are built from calcium carbonate. The reasons for this significant difference are not known. In both cases, however, the inorganic mineral is intimately associated with a complex assemblage of organic macromolecules—the *organic matrix*—that is of fundamental importance.

Calcium carbonate and calcium phosphate minerals have high lattice energies and low solubilities, and are therefore thermodynamically stable within biological environments. In contrast, hydrated phases, such as calcium oxalate and calcium sulfate, are much more soluble and therefore less common. In general, the precipitation of calcium salts provides an effective means to control the Ca^{2+} ion concentration in biological fluids. This helps to maintain a steady-state condition (*homeostasis*) corresponding to an intracellular calcium concentration of around 10^{-7} M.

Over half of the elements essential for life are incorporated in biomineral deposits. Of these, calcium is distinguished in being widespread and the common constituent of bones, teeth and shells.

2.1 Calcium carbonate—calcite and aragonite

There are six calcium carbonate minerals with the same principal composition but different structure; calcite, aragonite, vaterite, calcium carbonate monohydrate, calcium carbonate hexahydrate and amorphous calcium carbonate. Of these *polymorphs*, only the two most thermodynamically stable structures—*calcite and aragonite*—are deposited extensively as biominerals (Table 2.1). Magnesium ions are readily accommodated in the calcite lattice so many biological calcites also contain Mg^{2+} ions up to levels of 30 mol%.

2.1.1 Shells—big and small

Although molluscs build shells with all sorts of shapes and sizes they are generally conservative when it comes to the choice of mineral used. Except for a few oddballs—hydroxyapatite in the *Lingula* shell for example—shells are made pure and simply of calcium carbonate usually in the form of calcite and aragonite. Interestingly, many types of seashell contain both calcite and arag-

Table 2.1 Calcium carbonate biominerals

Mineral	Formula	Organism	Location	Function
Calcite	$CaCO_3$	Coccolithophores	Cell wall scales	Exoskeleton
		Foraminifera	Shell	Exoskeleton
		Trilobites	Eye lens	Optical imaging
		Molluscs	Shell	Exoskeleton
		Crustaceans	Crab cuticle	Mechanical strength
		Birds	Eggshells	Protection
		Mammals	Inner ear	Gravity receptor
Mg-calcite	$(Mg,Ca)CO_3$	Octocorals	Spicules	Mechanical strength
		Echinoderms	Shell/spines	Strength/protection
Aragonite	$CaCO_3$	Scleractinian corals	Cell wall	Exoskeleton
		Molluscs	Shell	Exoskeleton
		Gastropods	Love dart	Reproduction
		Cephalopods	Shell	Buoyancy device
		Fish	Head	Gravity receptor
Vaterite	$CaCO_3$	Gastropods	Shell	Exoskeleton
		Ascidians	Spicules	Protection
Amorphous	$CaCO_3 \cdot nH_2O$	Crustaceans	Crab cuticle	Mechanical strength
		Plants	Leaves	Calcium store

onite but the minerals are spatially separated in distinct parts of the shell. Usually the outer layer (*prismatic layer*) of the shell, as well as the growing edge, consists of large crystals of calcite, whereas the inner region (*nacre*) is built from a 'brick wall' of plate-like aragonite crystals (Fig. 2.1). The prismatic layer is first deposited and then the nacre is added with time as the shell grows in thickness. The remarkable switching between the calcium carbonate polymorphs is controlled by a layer of closely packed cells called the *outer epithelium* that is separated from the inner shell surface by a space filled with an aqueous solution—the *extrapallial space and fluid*, respectively

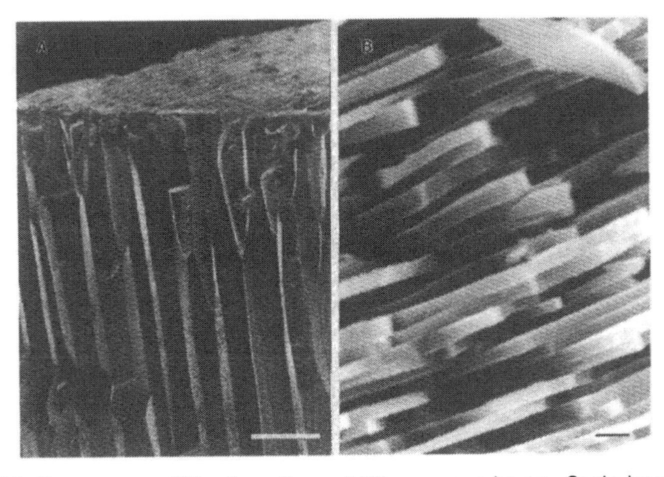

Fig. 2.1 Shell structures: (A) prismatic, and (B) nacreous layers. Scale bars, 50 and 1 μm, respectively.

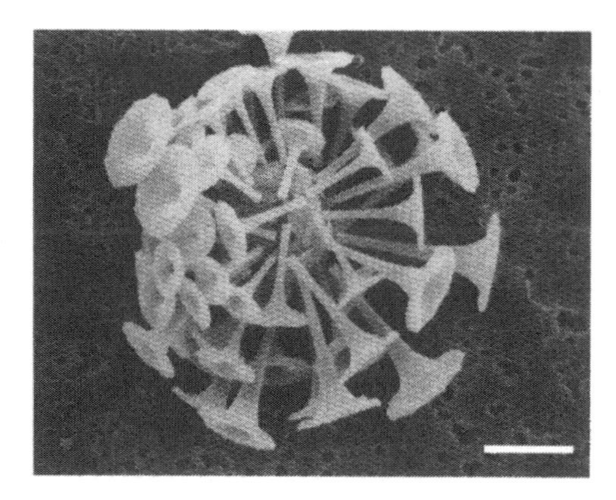

Fig. 2.3 Intact coccosphere with complex form. Scale bar, 3 μm.

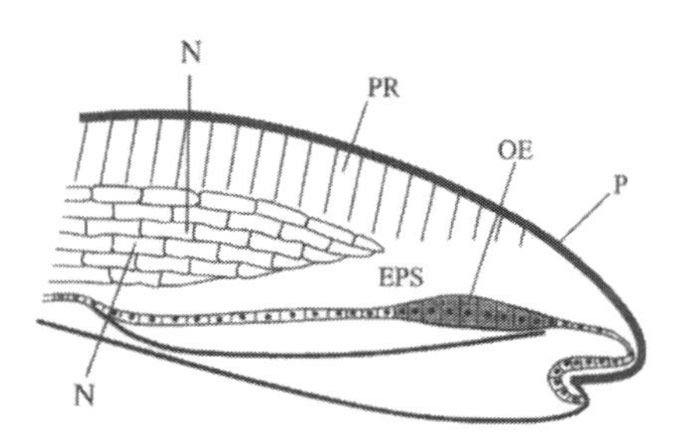

Fig. 2.2 Radial section of a shell showing positions of external organic layer (periostracum, P), prismatic mineral layer (PR), nacreous mineral layer (N), extrapallial space and fluid (EPS) and cells of the outer epithelium (OE).

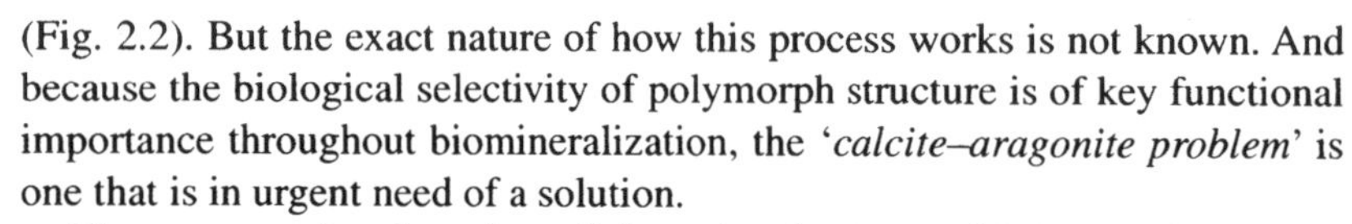

(Fig. 2.2). But the exact nature of how this process works is not known. And because the biological selectivity of polymorph structure is of key functional importance throughout biomineralization, the '*calcite–aragonite problem*' is one that is in urgent need of a solution.

The nacreous (mother-of-pearl) layer is a laminate of 0.5-μm-thick aragonite polygonal tablets sandwiched between thin (approximately 30 nm) sheets of a protein–polysaccharide organic matrix. The matrix plays a key role in limiting the thickness of the crystals and is structurally important in the mechanical design of the shell. It reduces the number of voids in the shell wall and so inhibits crack propagation by dissipating the energy associated with an expanding defect along the organic layers rather than through the inorganic crystals. This makes nacre approximately 3000 times as tough as inorganic aragonite! Further details of shell mineralization are described in Chapter 6, Section 6.5.

Some shell structures are very small indeed. Many marine single-celled organisms produce elaborate externally mineralized structures that serve as an 'exo-skeleton' in which the organism lives. Many of the most fascinating structures are found in a group of marine algae commonly known as *coccolithophores*. The mineralized shells (*coccospheres*) consist of wonderfully sculpted calcite plates called *coccoliths*, with additional decorative features such as long trumpet-shaped spines (Fig. 2.3). These structures are discussed in detail in Chapter 8, Section 8.4.1.

2.1.2 Gravity sensors

Calcite and aragonite are also used as *gravity sensors* in land and sea animals. These devices (generally referred to as *statoliths*, *statoconia*, *otoliths* or *otoconia*) function in a similar way to the fluid in the semicircular canals (which detect changes in angular momentum). In the human ear, the crystals are made of calcite and are spindle-shaped (Fig. 2.4), and sited on a specialized membrane under which sensory cells are located. During a change in linear acceleration, the movement of the crystal mass relative to the delicate hair-like extensions of the cells results in the electrical signalling of the applied

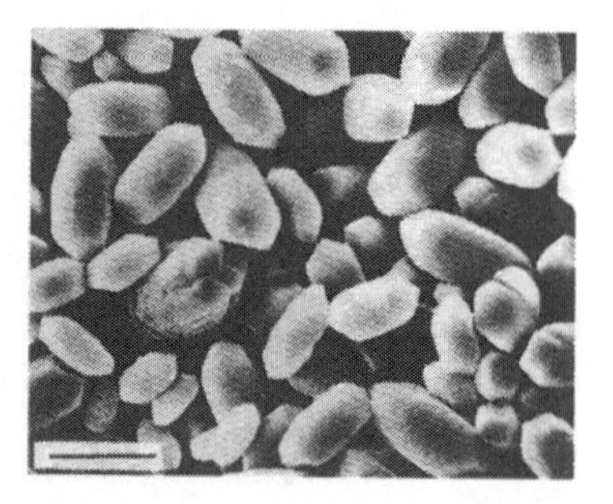

Fig. 2.4 Calcite crystals in the human inner ear. Scale bar, 8 μm.

force to the brain. This is why a cat lands on its feet if it falls out of a window—but don't try this on the neighbour's pet!

2.1.3 Lenses

Another use of calcite is as a *lens* in the compound eyes of creatures called *trilobites* (Fig. 2.5). These animals, which look like giant woodlice, are preserved as fossils, and the structure and organization of the corneal lenses have been determined by X-ray diffraction. The eyes consist of hexagonally packed arrays of calcite single crystals. Single crystals of calcite are well known for their ability to doubly refract white light, suggesting that the trilobites suffered a life of continual double vision (and ultimate extinction)! However, studies of well-preserved fossilized material show that each crystal is aligned in the eye such that the unique (non-refracting) crystallographic *c* axis is perpendicular to the surface of each lens. In this orientation, the calcite lens behaves isotropically like glass and a single well-defined image is formed. This nicely illustrates how the evolutionary design of biominerals is based on the interdependence of structure, organization and function.

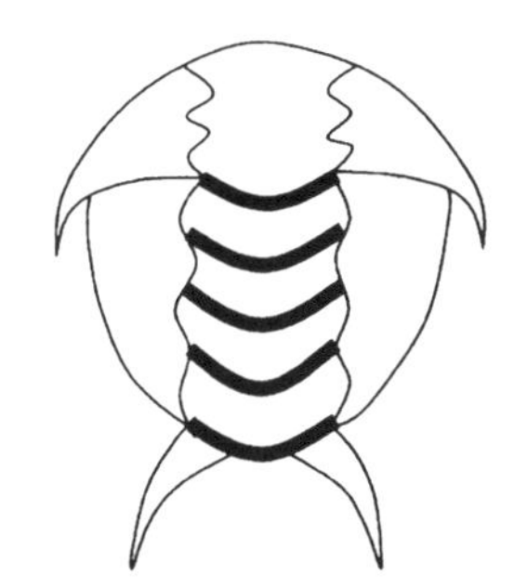

Fig. 2.5 Trilobite.

2.2 Calcium carbonate—vaterite and amorphous phases

Although most of the calcium carbonates formed in biological systems have structures of calcite or aragonite, some organisms deposit *vaterite*. Vaterite is the least thermodynamically stable of the three non-hydrated crystalline polymorphs and rapidly transforms to calcite or aragonite in aqueous solution. It occurs as elaborately shaped spicules in marine creatures called *ascidians* (the majority of calcareous sponges have magnesium-rich calcite spines), where it possibly acts as a structural support or as a deterrent against predators (Fig. 2.6). Vaterite has also been observed in the inner ears of two species of fish.

Amorphous calcium carbonate is formed in the leaves of many plants as spindle-shaped deposits (*cystoliths*) that act as a store of calcium (Fig. 2.7). Although this material is exceedingly unstable in inorganic systems due to

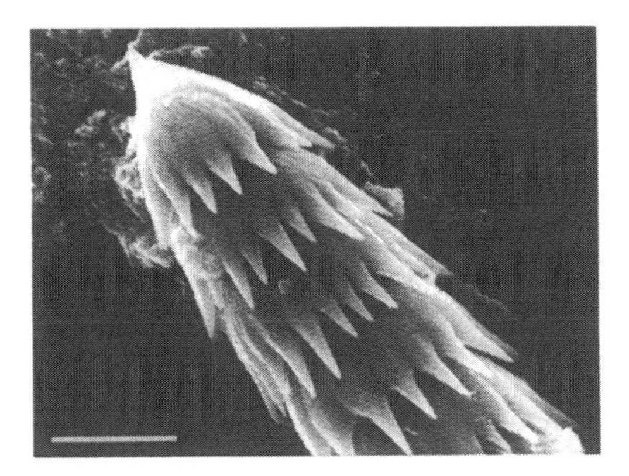

Fig. 2.6 Vaterite spicule. Scale bar, 5 μm.

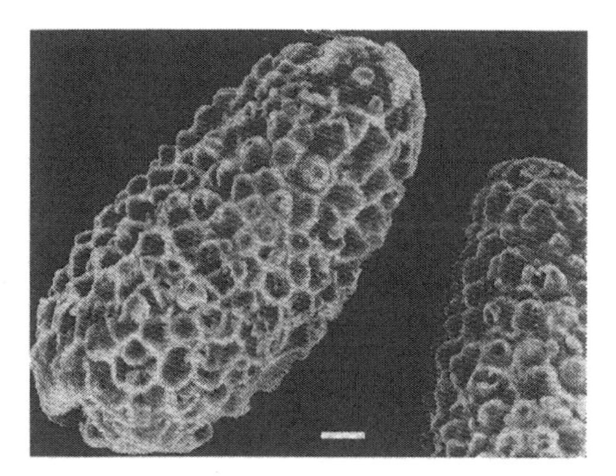

Fig. 2.7 Plant cystoliths. Scale bar, 10 μm.

rapid phase transformation in aqueous solution, the biomineral appears to be stabilized through the adsorption of biological macromolecules such as polysaccharides at the solid surface.

2.3 Calcium phosphate

Bone and *teeth* are made from calcium phosphate in the form of the mineral *hydroxyapatite* (HAP), along with a large number of proteins. The structural chemistry of biological hydroxyapatite is very complex because the mineral is not compositionally pure (non-stoichiometric), often being calcium-deficient and enriched in CO_3^{2-}, which replaces PO_4^{3-} ions in various lattice sites. Although in this book we will refer to bone mineral as hydroxyapatite, it is often known as 'carbonated apatite'. The composition can be expressed as:

$$(Ca,Sr,Mg,Na,H_2O,[\ ])_{10}(PO_4,HPO_4,CO_3,P_2O_7)_6(OH,F,Cl,H_2O,O,[\ ])_2$$

where [ ] denotes the presence of lattice defects. For most purposes, $Ca_{10}(PO_4)_6(OH)_2$ will suffice.

Several other calcium phosphate phases have been identified as intermediates in the biomineralization of calcium phosphates (Table 2.2). In particular,

Table 2.2 Calcium phosphate biominerals

Mineral	Formula	Organism	Location	Function
Hydroxyapatite	$Ca_{10}(PO_4)_6(OH)_2$	Vertebrates	Bone	Endoskeleton
		Mammals	Teeth	Cutting/grinding
		Fish	Scales	Protection
Octacalcium phosphate	$Ca_8H_2(PO_4)_6$	Vertebrates	Bone/teeth	Precursor phase
Amorphous	variable	Chitons	Teeth	Precursor phase
		Gastropods	Gizzard plates	Crushing
		Bivalves	Gills	Ion store
		Mammals	Mitochondria	Ion store
		Mammals	Milk	Ion store

there is evidence for an amorphous calcium phosphate phase in the early stage of bone and cartilage mineralization. Another phase, octacalcium phosphate ($Ca_8H_2(PO_4)_6$), has also been identified in various tissues where it readily transforms to HAP because of a close structural match between the unit cells of the two mineral phases.

2.3.1 Bone

Bone comes in all sorts of shapes and sizes in order to achieve the various functions of protection and mechanical support without compromising the requirement for mobility (Fig. 2.8). More than any other biomineral, the nature of bone highlights the important distinction between the inorganic and bioinorganic material world. Indeed, bone is often thought of as a 'living mineral' since it undergoes continual growth, dissolution and remodelling in response to both internal signals (during pregnancy for example) and external force fields, such a gravity.

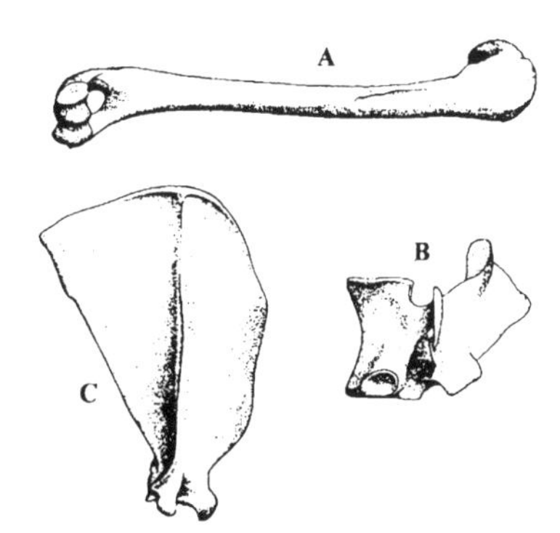

Fig. 2.8 Bone types: (A) long bone; (B) short bone; (C) flat bone.

The mechanical properties of bone are derived from the organized mineralization of hydroxyapatite within a matrix of *collagen fibrils*, glycoproteins (proteins with sugar side chains) and many other types of protein (see Chapter 6, Section 6.3, for more details). The combination of inorganic and organic components provides an increased toughness compared with hydroxyapatite alone, and this helps a lot if you want to go parachute jumping. By sculpting these components into micro-anatomical structures (Fig. 2.9)—woven bone, cortical bone, etc. (see Chapter 8, Section 8.1)—and controlling the amounts of mineral content, different levels of stiffness (referred to as *Young's modulus*) can be introduced into different bones according to their particular functions. A fast moving, highly agile animal such as a deer requires bones with high elasticity and relatively low mineral content (around 50 weight per cent). By contrast, the bones of a large marine mammal like the whale are stiff, with a hydroxyapatite content greater than 80 wt% (Fig. 2.10).

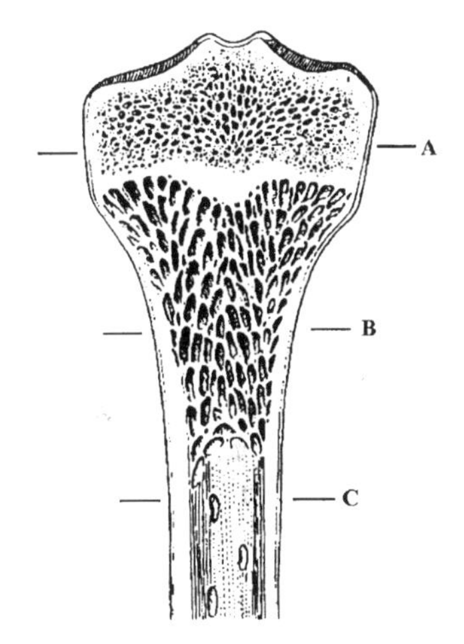

Fig. 2.9 Internal structure of long bone with three different microstructures (A, B and C).

The non-stoichiometric nature of bone mineral may be responsible for the apparent *piezoelectric response* observed in this tissue. Although the precise mechanisms are unknown, the application of pressure stimulates the growth of bone mineral. (To put this principle into a vivid historical context, soldiers of the First World War had their broken legs stimulated by a helpful matron wielding a wooden mallet!) Bone contains a network of cells that live within the mineralized structure and are interconnected through small pores and channels. One possibility is that these cells (*osteocytes*) act as biological 'strain gauges' that respond to changes in mechanical pressure and send chemical or electrochemical signals to the bone surface which then activate another type of cell called an *osteoblast* to begin mineralization. The process of activation is further complicated because there is another type of cell called an *osteoclast* whose job it is to degrade bone through a heady cocktail of acid and enzymes, and these also respond to the signalling. Overall, the process is incredibly complex and subject to many forms of feedback controlled by a large number of biochemical triggers such as hormones that are circulating in the blood stream.

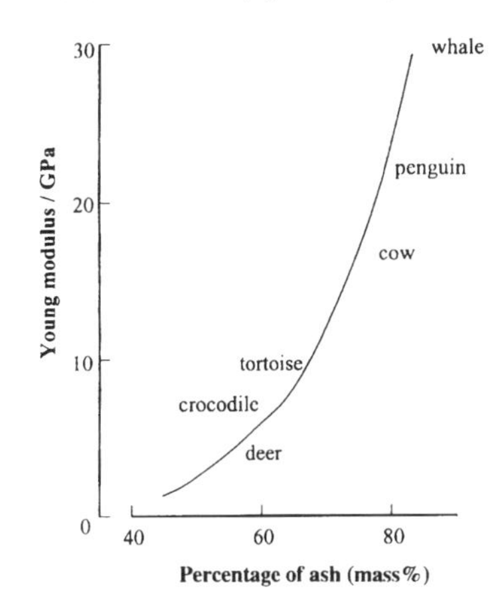

Fig. 2.10 Increase in stiffness (Young's modulus) with increasing mineral content (mass%) for various cortical bones.

2.3.2 Teeth

The structure and organization of *tooth enamel* and *dentine* (Fig. 2.11), like bone, derive from a highly complex system designed to withstand specific types of mechanical stress. Enamel, which is on the outside of the tooth, is much less tough than bone because it is close to 95 per cent by weight hydroxyapatite (human bone on average is around 65 per cent) but gains some structural resistance by interweaving long ribbon-like crystals into an inorganic fabric (Fig. 2.12). Interestingly, enamel starts out with a high proportion of proteins (mainly *amelogenin* and *enamelin*, see Chapter 6, Section 6.4) which are progressively removed as the biomineral matures to produce the high mineral volume fraction of the erupted tooth (Fig. 2.13). *Dentine*, on the other hand, which resides within the central regions of the tooth, contains collagen and is more similar in structure and composition to bone.

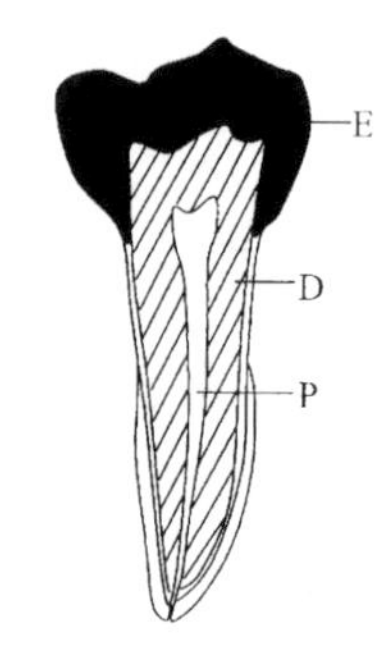

Fig. 2.11 Human tooth with outer enamel (E), inner dentine (D) and pulp (P).

A principal cause of the general increase in dental health in many societies is the use of fluoride in drinking water and in numerous toothpastes. The F^- ion is readily incorporated into the hydroxyapatite lattice where it stabilizes the lattice and reduces the solubility of the mineral phase. Interestingly, fish teeth consist of a structure very similar to enamel (called *enameloid*) but which contains high levels of natural fluoride. For example, the fluoride concentration in shark enameloid is over one thousand times that in human enamel (Table 2.3).

Fig. 2.12 The internal structure of enamel. Scale bar, 1 μm.

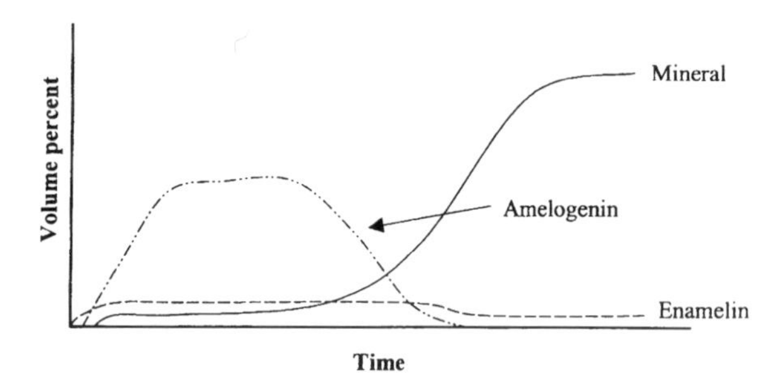

Fig. 2.13 Loss of proteins and increase in mineral content with time during enamel formation.

Table 2.3 Chemical composition of calcium phosphate (hydroxyapatite) in human and shark enamel

Composition (wt%)	Human enamel	Shark enamel
Ca^{2+}	37.55	37.26
Na^{+}	0.75	0.76
Mg^{2+}	0.27	0.32
PO_4^{3-}	17.68	17.91
CO_3^{2-}	3.6	1.1
F^-	0.02	3.65

Table 2.4 Group 2A sulfate and oxalate biominerals

Mineral	Formula	Organism	Location	Function
Gypsum	$CaSO_4 \cdot 2H_2O$	Jellyfish	Statoconia	Gravity receptor
Celestite	$SrSO_4$	Acantharia	Cellular	Micro-skeleton
Barite	$BaSO_4$	Loxedes	Intracellular	Gravity receptor
		Xenophyophores	Intracellular	Unknown
		Chara	Statoliths	Gravity receptor
Whewellite	$CaC_2O_4 \cdot H_2O$	Plants/fungi	Leaves/roots	Calcium store
Weddellite	$CaC_2O_4 \cdot 2H_2O$	Plants/fungi	Leaves/roots	Calcium store

2.4 Other Group 2A biominerals

Remarkably, some unicellular organisms control the deposition of Group 2A sulfate minerals for specific functions (Table 2.4). *Barite* in particular has a high specific gravity and therefore makes a good candidate for sensing gravity (Fig. 2.14). In unicellular organisms called *desmids*, the crystals are often located at either ends of the elongated cell so that the two ends bend to give a crescent-like morphology. In *Loxedes*, the barite crystals are contained within a small spherical bag that is attached to the end of a filamentous structure to produce a biological 'ballcock' that senses changes in orientation.

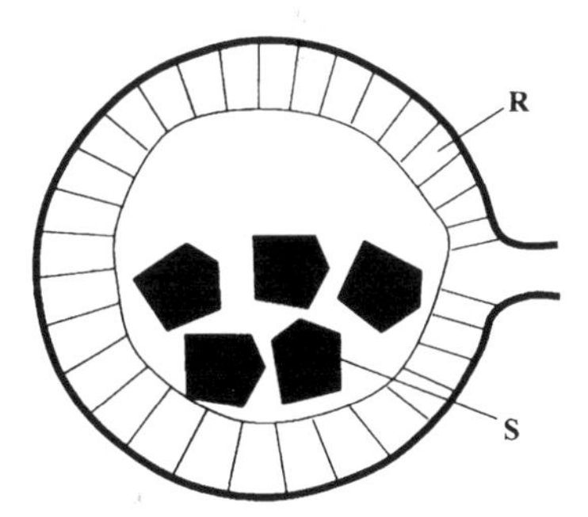

Fig. 2.14 Gravity sensor based on mineral grains (statoconia, S) and surrounding receptor cells (R).

The micro-skeletal structures of *acantharians* are constructed from interlocking spines of *celestite*. Most species build structures containing 20 spines, each of which is a single crystal oriented with the crystallographic *a* axis parallel to the morphological long axis. The spicules are arranged in a highly regular fashion such that the overall structure has D_{4h} point group symmetry (Fig. 2.15). This means that a series of imaginary straight lines drawn between the tips of the spines would encase the structure in a rectangular box with different dimensions for the length, width and height.

Calcium oxalates are also found as hydrated minerals in many plants (Table 2.4).

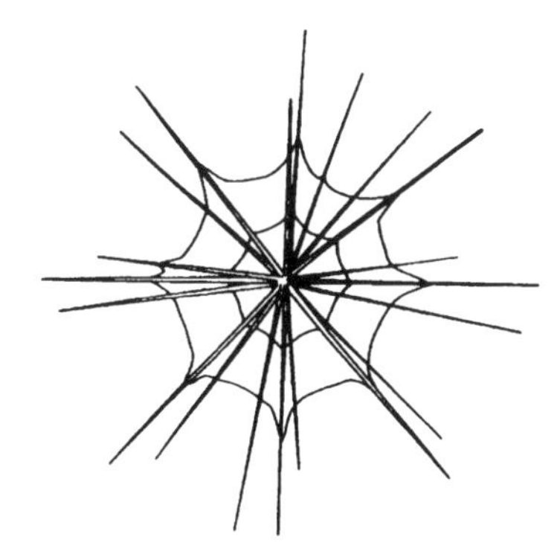

Fig. 2.15 Acantharian skeleton with $SrSO_4$ spines.

2.5 Silica

Although most biominerals are ionic salts, many unicellular organisms produce remarkable structures from *amorphous silica* (Table 2.5). Among these, *diatoms* (Fig. 2.16) and *radiolarians* (Fig. 2.17) are famous for their medusa-like propensity to construct delicate lace-like porous shells (*frustules*) and micro-skeletons, respectively. How these complex shapes are produced are described in Chapter 7, Section 7.3.2.

Why some organisms utilize amorphous silica rather than a crystalline mineral such as calcium carbonate as a structural material is unknown. One possibility is that because the fracture and cleavage planes inherent to crystalline structures are missing, the amorphous biomineral can be subsequently moulded without loss of strength into a wide variety of complex architec-

Table 2.5 Silica biominerals

Mineral	Formula	Organism	Location	Function
Silica	$SiO_2 \cdot nH_2O$	Diatoms	Cell wall	Exoskeletons
		Choanoflagellates	Cellular	Protection
		Radiolarians	Cellular	Micro-skeleton
		Chrysophyts	Cell wall scales	Protection
		Limpets	Teeth	Grinding
		Plants	Leaves	Protection

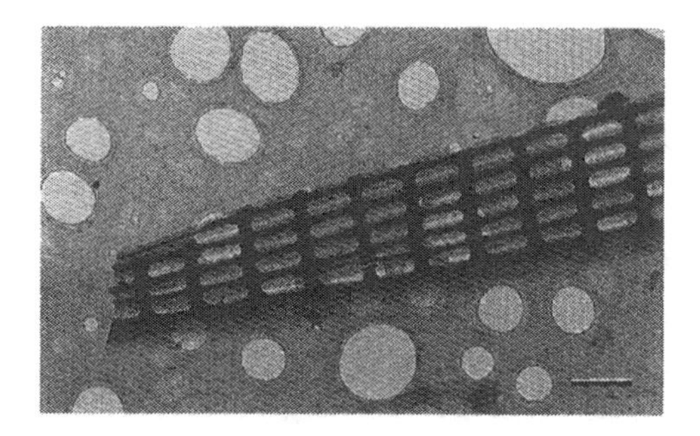

Fig. 2.16 Diatom shell (frustule). Scale bar, 1 μm.

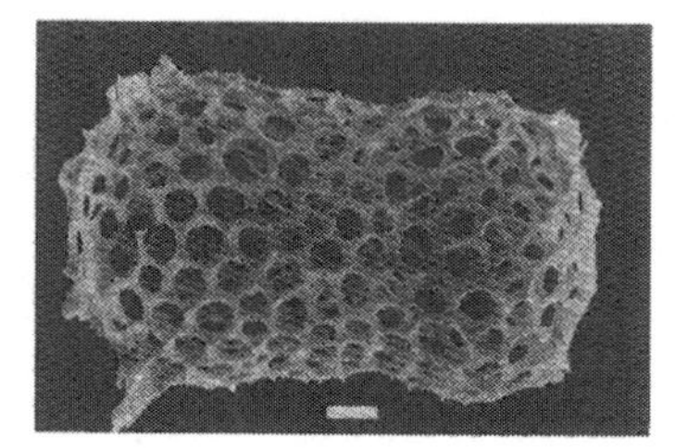

Fig. 2.17 Radiolarian micro-skeleton. Scale bar, 10 μm.

tures. In higher plants, the presence of a large number of silica spines and nodules (*phytoliths*) in the cell walls serves up an unpalatable meal for a discerning predator, as well as producing skid marks on sports day. Plants such as *horsetails* can have silica levels of 20 to 25 per cent of the dry weight, so high in fact that the early American pioneers used the plant as an effective, if not abrasive, means of cleaning teeth! Similar high levels are found in rice husks, which have been used as a cheap raw material for the industrial preparation of silicon nitride.

The high stability of Si–O–Si units in water and the variability allowed in the Si–O–Si bond angle are responsible for the precipitation of the disordered rather than crystalline structure under ambient conditions. Much higher temperatures are therefore required for the crystallization of quartz—sand, unlike chalk, is not a biogenic product. There are also other significant differences between the deposition of silica and crystallization of ionic minerals, such as calcium carbonate and phosphate (see Chapter 4 for a discussion of crystallization). At neutral pH, the soluble form of silica is silicic acid, $Si(OH)_4$, which is a weak acid with a pK_a of 9.8. Above concentrations of 1 mM, silicic acid undergoes a series of polycondensation reactions to produce amorphous gels or colloidal particles. However, as seawater is generally undersaturated with respect to silica deposition, organisms have to employ specific mechanisms for the active uptake, concentration and polycondensation of silicic acid from the environment. The resulting amorphous silica has a structure based on a covalently linked polymeric network of randomly arranged tetrahedrally coordinated siloxane centres with variable levels of hydroxylation (Fig. 2.18). These range from fully condensed centres ($-O_3SiO-$, so-called Q_4

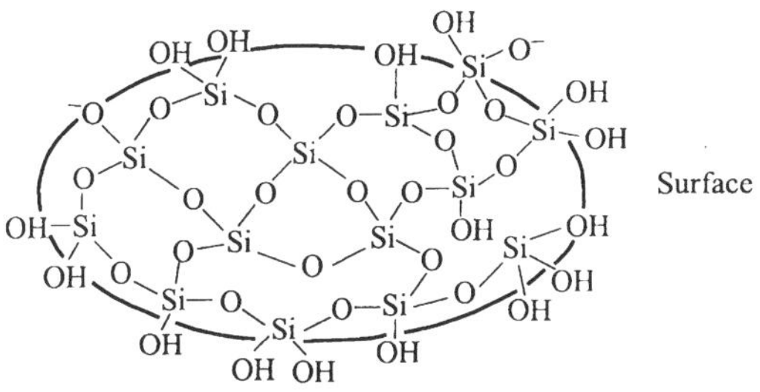

Fig. 2.18 The chemical nature of amorphous silica.

sites), to partially condensed $-O_3SiOH$ (Q_3), $-O_2Si(OH)_2$ (Q_2) and $-OSi(OH)_3$ (Q_1) centres to give a complex hydrated material with general composition ($[SiO_{n/2}(OH)_{4-n}]_m$, $n = 1$ to 4). The relative proportion of these centres is determined by solid-state ^{29}Si NMR spectroscopy and can vary in different silica biominerals (Fig. 2.19).

Although amorphous at the molecular level, silica biominerals adopt specific microscopic textures such as gels or particulate structures consisting of closely packed aggregates of colloidal particles. These motifs reflect differences in the extent of aggregation and fusion of small primary particles that form initially as the polycondensation process gets underway. In pure solution at neutral pH and above, the primary particles are slightly negatively charged, so the extent of aggregation is limited by the build-up of electrostatic charge, and a particulate structure is produced. The precise arrangement of the constituent particles depends on the presence of an organic matrix, and this directs the aggregation process along specific directions. For example, seeds of the canary grass are covered in fine silica hairs that contain three distinct silica motifs—fibrillar, sheet-like and granular—in conjunction with different macromolecules (Fig. 2.20). The gel texture, on the other hand, arises by screening the surface charge of the primary particles by cations and positively charged polyelectrolytes present in the surrounding environment.

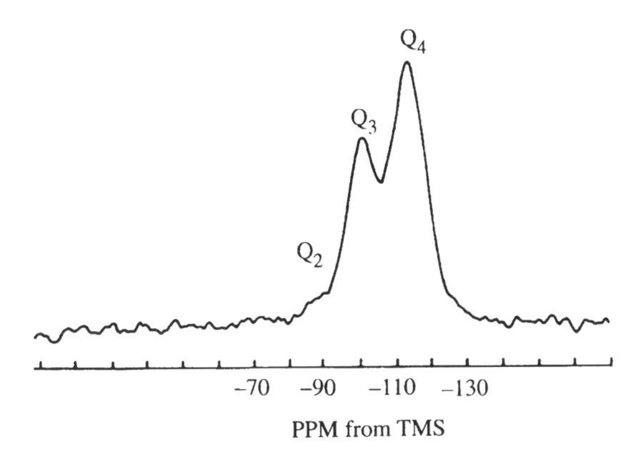

Fig. 2.19 ^{29}Si NMR spectrum for silica sponge spicules. See text for peak assignments. TMS, tetramethylsilane.

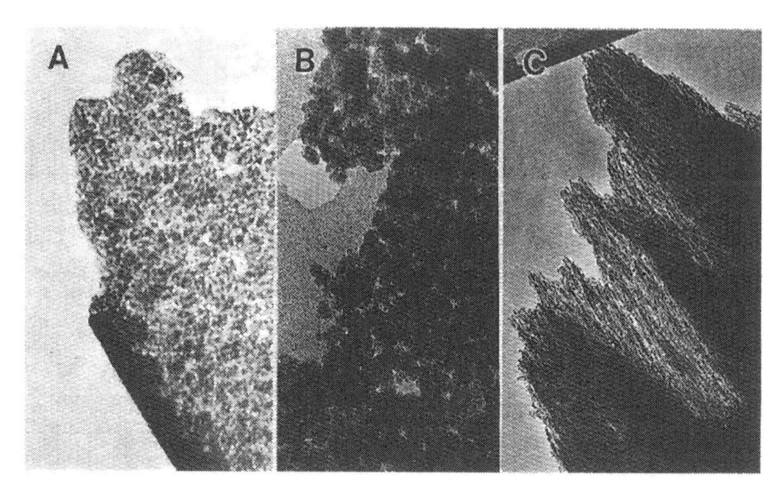

Fig. 2.20 Types of plant silica: (A) sheet-like; (B) globular; (C) fibrillar.

2.6 Iron oxides

Bioinorganic iron oxides are widespread and serve several different functions (Table 2.6). The minerals have important counterparts in inorganic chemistry, where they are extensively used in catalytic and magnetic devices.

2.6.1 Magnetic bacteria

In rivers and ponds throughout the world there are amazing bugs that navigate in the earth's magnetic field. These *magnetotactic bacteria* have an internal compass needle that consists of a chain of discrete crystals of the mixed valence compound *magnetite* (Fe_3O_4 or $Fe^{III}_2Fe^{II}O_4$) (Fig. 2.21). Together, the crystals align the organism along the lines of force of the geomagnetic field. Unless you are at the north or south pole, the earth's field has both a vertical and horizontal component so this means that magnetic bacteria in the

Table 2.6 Iron oxide biominerals

Mineral	Formula	Organism	Location	Function
Magnetite	Fe_3O_4	Bacteria	Intracellular	Magnetotaxis
		Chitons	Teeth	Grinding
		Tuna/salmon	Head	Magnetic navigation
Geothite	α-FeOOH	Limpets	Teeth	Grinding
Lepidocrocite	γ-FeOOH	Sponges	Filaments	Unknown
		Chitons	Teeth	Grinding
Ferrihydrite	$5Fe_2O_3{\cdot}9H_2O$	Animals/plants	Ferritin	Storage protein
		Chitons	Teeth	Precursor phase
		Beaver/rat/fish	Tooth surface	Mechanical strength
	+ phosphate	Bacteria	Ferritin	Storage protein
		Sea cucumber	Dermis	Mechanical strength

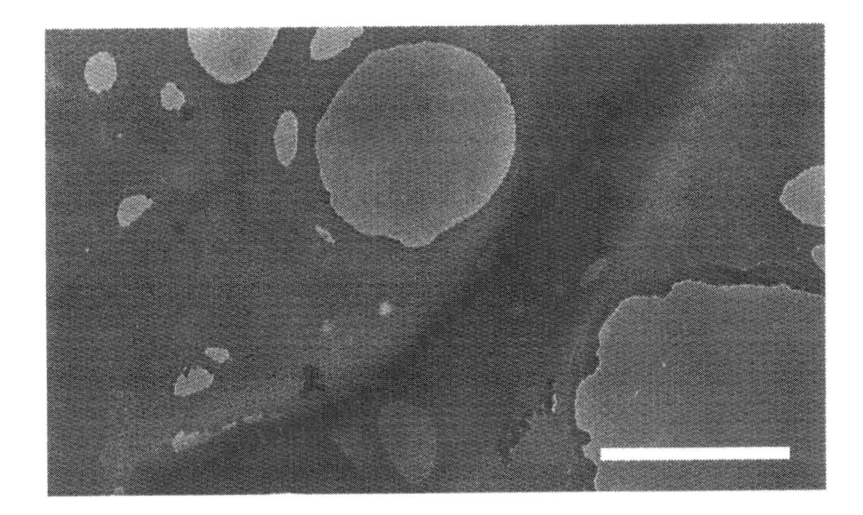

Fig. 2.21 Linear chain of magnetite crystals in a magnetotactic bacterium. Scale bar, 500 nm.

Northern hemisphere can navigate in a downwards direction only if the biological compass needle is north-seeking (Fig. 2.22). Because many of these organisms live at the sediment–water interface, having a downward-seeking magnetotaxis provides a quick and reliable means of finding this zone in turbulent conditions. In fact, the magnetic bacteria zoom along in a straight helical flight path whereas non-magnetic bugs undertake a 'run-and-tumble' procedure, sniffing out the local chemical conditions and then randomly sampling the next location, like a drunkard's walk on a pub crawl. Scientists have fished for these magnetotactic bacteria in freshwater and marine environments around the globe (you need a small bar magnet as bait!). Significantly, the same species found in the Northern hemisphere exist also in the Southern hemisphere and have a similar arrangement of magnetite crystals but of the opposite polarity—that is, the cells are south-seeking, so they also swim exclusively downwards.

In each known example of magnetotactic bacteria, the magnetite crystals are not only aligned in chains but also have dimensions compatible with those required for a single magnetic domain. Otherwise the system would not function efficiently. In the bulk state, magnetite is classified as a *ferrimagnetic*

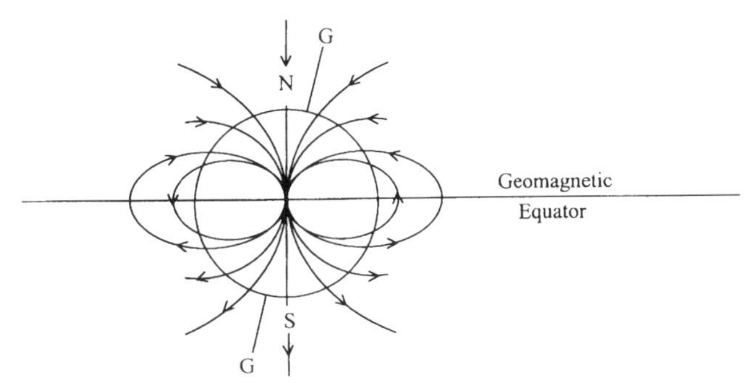

Fig. 2.22 The earth's magnetic field. N, north; S, south; G, geographical poles.

oxide, which means that the permanent magnetic dipole moment arises from the non-zero sum of two opposing magnetic sublattices (ferromagnetism in contrast has all the atomic spins aligned in the same way). However, the magnetism of individual crystals depends critically on their size and shape. For example, if they are cubic and smaller than about 5 nm then the particles are too small to retain the permanent magnetic dipole because the thermal energy keeps flipping the aligned atomic spins along different directions of the crystal lattice. This is because the large surface-to-area ratio of such small particles gives rise to a large number of iron atoms in relatively unstable surface sites. The thermal scrambling of the magnetic dipole—an effect referred to as *superparamagnetism*—can be offset if the temperature is reduced, or if a strong applied field is introduced to marshal the spins into one specific direction. Alternatively, if the particle size becomes larger than 5 nm (or the shape changes to an elongated prism), the magnetic field adopts a single permanent direction (called the easy axis of magnetization) because the relative number of surface verses bulk atoms decreases. This is the optimum condition with the greatest magnetism per unit volume because if the cubic-shaped particles become even larger, for example greater than 10 nm, the single domain particles separate internally into several antiparallel magnetic domains (Fig. 2.23). Given the elegance of evolutionary design, it is perhaps not too surprising that this narrow-sized window has been exploited in bacterial magnetites which all fall within the size restrictions for a permanent single magnetic domain.

2.6.2 Rusty proteins

An important and widespread iron oxide is the hydrated mineral called *ferrihydrite* ($5Fe_2O_3 \cdot 9H_2O$). This is the brown gelatinous precipitate which is readily formed in a test-tube by the addition of sodium hydroxide to an Fe^{III} solution. The iron storage protein *ferritin* contains approximately 30 wt% of iron in the form of a 5-nm ferrihydrite core surrounded by a polypeptide coat (Fig. 2.24). Each inorganic nanoparticle is synthesized inside a preformed empty polypeptide shell, referred to as *apoferritin*, by a specific chemical mechanism (see Chapter 5, Section 5.1.2, and Chapter 6, Section 6.7.3, for details). The process not only provides a storage depot for up to 4500 Fe atoms, but also produces a colloidal sol of non-aggregated nanoparticles (Fig. 2.25).

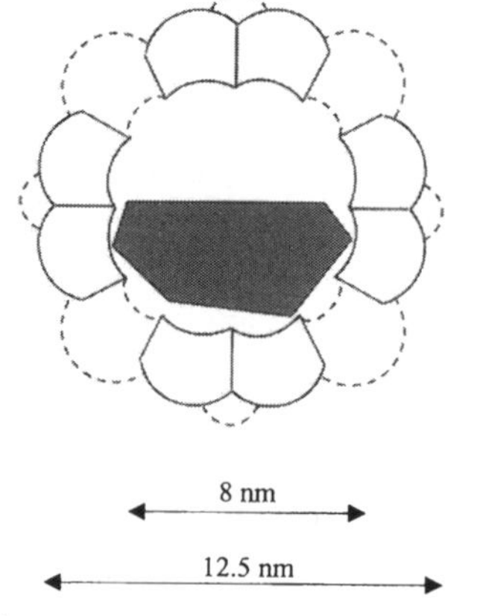

Fig. 2.24 Cross-section of the ferritin molecule with polypeptide shell and mineral core.

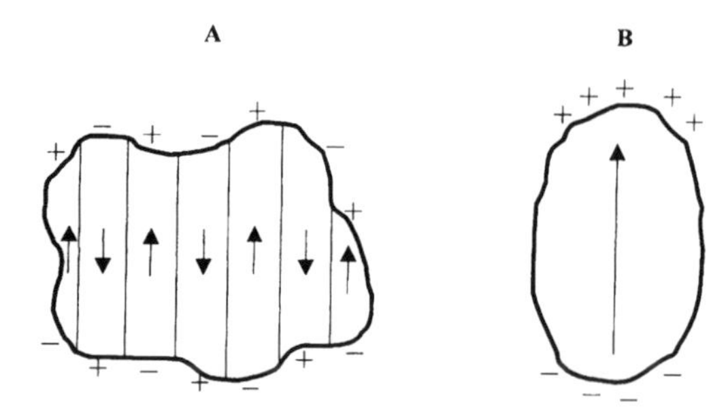

Fig. 2.23 Magnetite particles with (A) multidomain and (B) single-domain magnetic structures.

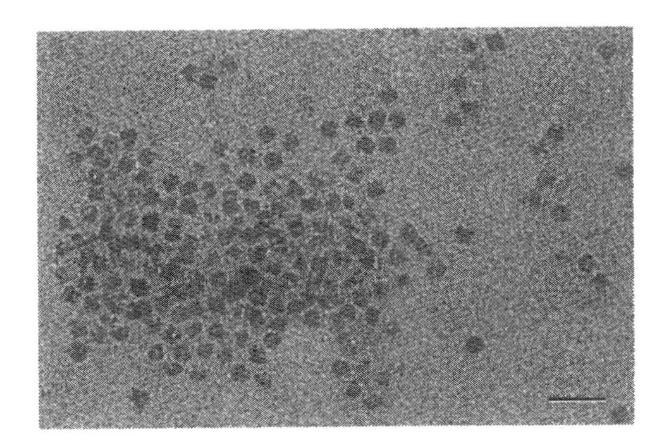

Fig. 2.25 Discrete nanoparticles of ferritin. Scale bar, 20 nm.

As a solubilized form of iron oxide, ferritin can be readily transported between different locations without the organism rusting up! There are, however, several major diseases, such as β-thalassemia, that are associated with the storage and transport of iron and the pathology of 'iron overload'. In general, ferritin protects the cells from labile iron by sequestration and acts as a buffer for regulating the levels of free iron. The latter is possible because of the relative high solubility of ferrihydrite at slightly acidic pH values, which allows iron to be not only stored at neutral pH but also released at pH levels of about 5.5. This takes place in cellular structures called *lysosomes* and the released iron is used for biochemical processes such as the synthesis of haemoglobin. Other iron oxides, for example magnetite, are much less soluble because of their greater thermodynamic stability, and do not function as reversible storage sites in biology.

Unlike magnetite, biologically produced ferrihydrite is variable in both structure and composition. For example, the mineral cores of horse spleen ferritin—which is commercially available in large amounts—are relatively well-ordered single crystals and contain a small amount of inorganic phosphate (usually about 5 mol% of the Fe content). In contrast, ferritins extracted from bacteria have amorphous cores and phosphate concentrations almost as high as for iron.

These differences in composition also account for the variable magnetic properties of the ferritin cores. Ferrihydrite is classified as an *antiferromagnetic* oxide in which the magnetic sublattices are perfectly opposed to give a zero net magnetic dipole moment. But as with magnetite (Section 2.6.1), the magnetism is dramatically influenced by size and shape. In fact, ferritin cores show *superparamagnetic* behaviour due to an excess of uncompensated spins that originate from the relatively large number of coordinatively unsaturated surface atoms. This is most clearly seen by studying samples at different temperatures using a ^{57}Fe Mössbauer spectrometer. This technique uses gamma rays to excite the iron nucleus and the energy absorbed in this process depends on the oxidation state, spin state and coordination geometry of the iron atoms, as well as the magnetic field around the nuclei. For mammalian ferritin, the spectra show a doublet at temperatures above 40 K, a sextet at 4.2 K and below (Mössbauer spectroscopists like to do it in the cold), and a mixture of the doublet and sextet in between (Fig. 2.26). The six-line spec-

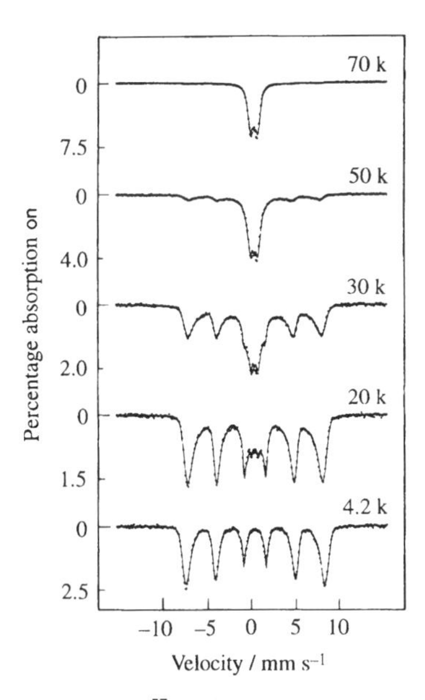

Fig. 2.26 ^{57}Fe Mössbauer spectra for mammalian ferritin at different temperatures.

trum is a classic signature for magnetic ordering and arises because of the alignment of the uncompensated spins. In contrast, the two-line spectrum, which increases in intensity with temperature, indicates that the magnetic ordering begins to fluctuate rapidly as the thermal energy is increased—so fast in fact that individual magnetic states can no longer be picked up by the Mössbauer technique. When similar experiments are run with bacterial ferritin, no six-line spectrum is observed even at 1.3 K. This indicates that the high levels of phosphate and the concomitant amorphous nature of the ferritin nanoparticles are responsible for the absence of magnetic order in these proteins.

2.6.3 Iron teeth

Molluscs such as *limpets* and *chitons* have unusual dining habits. Although they don't appear to do much at low tide, once submerged they begin a frenzied feeding activity, scraping algae off the intertidal rocks with a tongue-like organ called the *radula*. In the common limpet (*Patella vulgata*), the radula is about 7 cm long, a few millimetres wide and contains hundreds of teeth (Fig. 2.27). One end of the radula protrudes from the mouth and a small number of fully formed teeth are rasped across the rock surface to loosen the algae and other microorganisms. Because the algae are cemented tightly to the rock, the teeth have to be very hard and tough. Both limpets and chitons solve this problem by arming themselves with sabre-like teeth that contain various crystalline iron oxide minerals in the cutting edge and backed with deposits of silica or hydroxyapatite, respectively. In the common limpet the mineral phase is *goethite* (α-FeOOH), whereas a mixture of *lepidocrocite* (β-FeOOH) and magnetite is deposited in chiton teeth (Fig. 2.28). Limpet and chiton teeth can therefore by readily distinguished because the latter are magnetic and easily picked up (without the attached organism!) by a small bar magnet.

Fig. 2.28 Cross-section of chiton tooth showing three minerals: magnetite cutting edge (black), lepidocrocite (hatched) and hydroxyapatite (white).

One of the amazing things about the radula is that it is a continuous conveyer belt of teeth at progressively different stages of mineralization (Fig. 2.29). Whereas the teeth exposed in the mouth are hard mineralized structures, those further back are less developed, and those at the far end are soft organic models covered in cells called *odontoblasts*. As the mature teeth

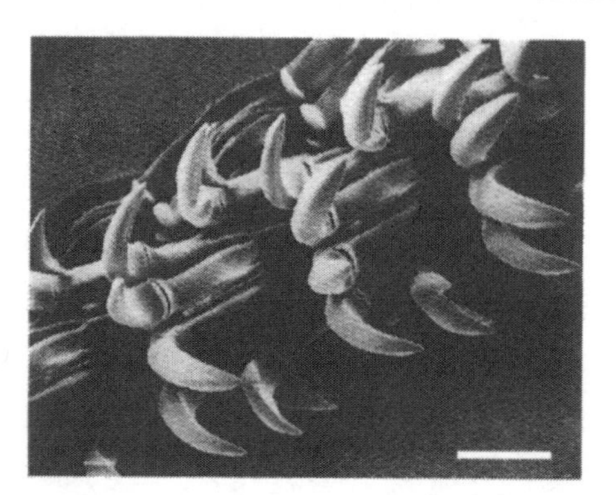

Fig. 2.27 Limpet teeth. Scale bar, 200 μm.

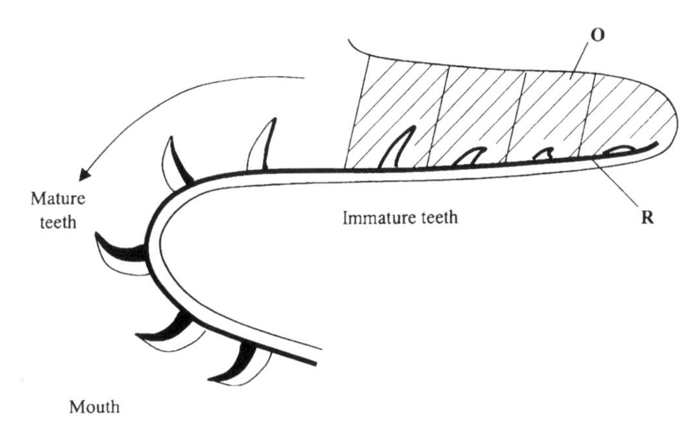

Fig. 2.29 Cross-section showing limpet/chiton teeth at different stages of development along radula (R). The initial stages are associated with odontoblast cells (O).

become abraded during feeding, they are replenished on a regular basis by forward movement of the radula. The development of the teeth along the radula can therefore be studied in a single organism. In general, the process is highly complex and involves the inclusion of secondary minerals, such as silica (limpets) or amorphous calcium phosphate (chitons), and changes in the organic matrix present within the tooth.

2.7 Metal sulfides

Many different types of iron sulfide minerals are formed in association with *sulfate-reducing bacteria* but they show none of the characteristics of a controlled biomineralization process. Most of these products are adventitious and arise from the reaction of metabolic products such as H_2S with Fe^{II} species in the surrounding environment. However, recent studies have shown that certain types of *magnetotactic bacteria* present in sulfide-rich environments synthesize and organize crystals of the ferrimagnetic mineral *greigite* (Fe_3S_4) (Fig. 2.30). The inclusions are discrete single crystals of narrow size distribution, have species-specific morphologies and appear to be crystallographically aligned within chains, just like the magnetite crystals described in Section 2.6.1. Because the early stages of the earth's history are generally associated with a reducing rather than an oxidizing environment, intracellular deposition of the bacterial iron sulfides may possibly reflect an ancient process that predates magnetite biomineralization by millions of years.

Interestingly, some *yeasts* have evolved the means to mineralize quantum dot semiconductors! Many plants and fungi contain a family of short chelating peptides called *phytochelatins* with the general structure $(\gamma\text{-EC})_n\text{-G}$, where E is glutamic acid, C is cysteine and G is glycine (Fig. 2.31). The linkage between the repeating glutamylcysteine dipeptide involves the γ-carboxyl rather than the more typical α-carboxyl of most proteins and peptides, as shown in Fig. 2.31. Exposure of yeasts to relatively high levels of Cd^{II} ions

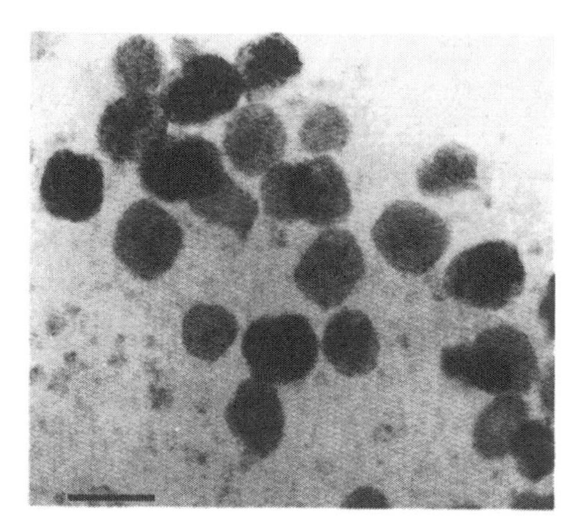

Fig. 2.30 Greigite (Fe_3S_4) crystals extracted from magnetotactic bacteria. Scale bar, 100 nm.

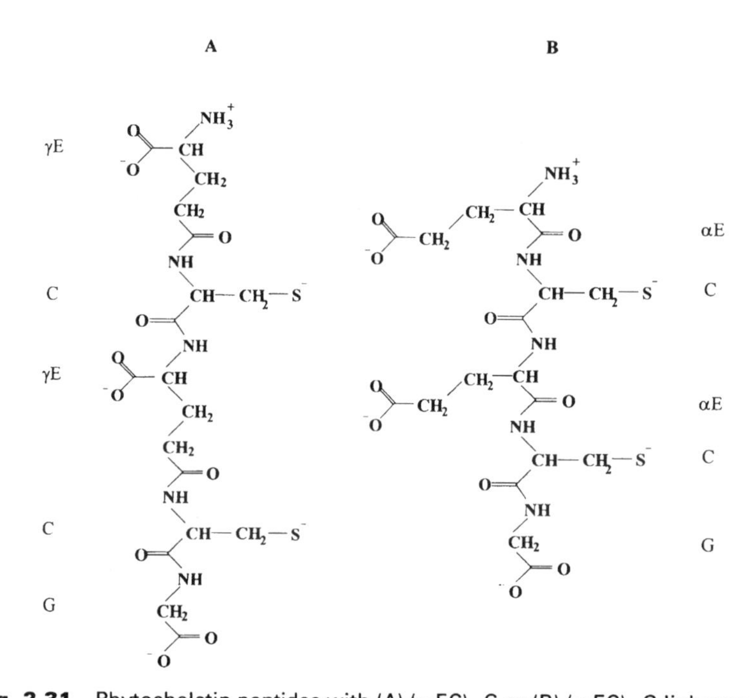

Fig. 2.31 Phytocheletin peptides with (A) (γ-EC)$_2$-G or (B) (α-EC)$_2$-G linkages.

results in the formation of a heterogeneous mixture of phytochelatins with two to four repeating units that bind a cluster of the metal ions. These then serve as nucleation sites for the formation of intracellular CdS nanoparticles. The size of the particles is restricted to only a few nanometres (approximately 85 CdS units) because the growth process is arrested by surface binding of the peptides to the inorganic clusters (Fig. 2.32).

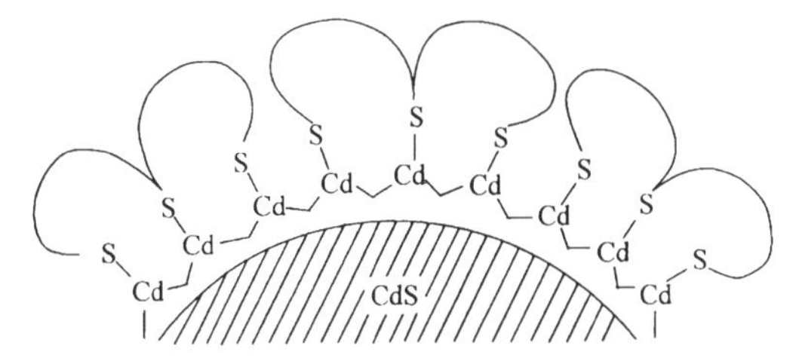

Fig. 2.32 CdS nanoparticle surface with anchored phytocheletin molecules.

2.8 Summary

In this chapter we have described how biominerals, such as calcium carbonate, calcium phosphate, amorphous silica and iron oxides, are deposited as functional materials in a wide range of organisms. Many of these biominerals have remarkable levels of complexity. In the next chapter, we sketch out some general principles that will begin to help us to understand how such exquisite structures can be produced.

Further reading

Bäuerlein, E. (2000). *Biomineralization: from biology to biotechnology and medical application*. Wiley-VCH, Weinheim.

Frankel, R. B. and Blakemore, R. P. (1991). *Iron biominerals*. Plenum Press, New York.

Leadbeater, B. S. C. and Riding, R. (1986). *Biomineralization in lower plants and animals*. Systematics Association Vol. 30. Oxford University Press, Oxford.

Lowenstam, H. A. and Weiner, S. (1989). *On biomineralization*. Oxford University Press, New York.

Mann, S., Webb, J. and Williams, R. J. P. (1989). *Biomineralization: chemical and biochemical perspectives*. VCH Verlagsgesellschaft, Weinheim.

Miller, A., Phillips, D. and Williams, R. J. P. (1984). Mineral phases in biology. *Philos. Trans. R. Soc. London B*, **304**, 409–588.

Simkiss, K. and Wilbur, K. (1989). *Biomineralization: cell biology and mineral deposition*. Academic Press, San Diego.

Volcani, B. E. and Simpson, T. L. (1982). *Silicon and siliceous structures in biological systems*. Springer Verlag, Berlin.

3 General principles of biomineralization

Although we know a lot about the structures of biominerals and how they vary in different organisms (see Chapter 2), relatively little is known in precise detail concerning the molecular interactions governing their controlled construction. There are, however, some general principles that are reasonably well established and which we discuss in this chapter. These principles are then developed further in Chapters 4 to 8 with a focus throughout on the prevailing ideas and concepts, particularly from a chemical perspective. Other important aspects, such as the genetic and cellular control of biomineralization, lie for the most part outside the scope of this book.

3.1 Biologically induced mineralization

In *biologically induced mineralization*, inorganic minerals are deposited by adventitious precipitation, which arises from secondary interactions between various metabolic processes and the surrounding environment. For example, in certain types of green algae, calcium carbonates are precipitated from saturated calcium bicarbonate solutions by metabolic removal of carbon dioxide during photosynthesis, according to the chemical equilibrium,

$$Ca^{2+} + 2HCO_3^- \rightleftharpoons CaCO_3 + CO_2 + H_2O$$

In a similar way, the extrusion of metabolic products across or into the cell wall of bacteria can result in the biologically induced precipitation of various inorganic minerals by subsequent reaction with extraneous metal ions (Table 3.1). In particular, fluxes of OH^- are involved with the deposition of oxides, carbonates and phosphates, whereas H_2S and electrons induce the precipitation of sulfides and mixed valance iron oxides (Fig. 3.1). Some bacteria even have the ability to accumulate and passivate toxic metal ions, such as UO_2^{2+}, Pb^{II} and Cd^{II}. The selective coprecipitation of these ions suggests that biologically induced mineralization could have an important role in aiding the clean up of polluted waters and soils.

One distinctive feature of biologically induced mineralization is that the minerals usually form along the surface of the cell where they remain firmly attached to the cell wall (Fig. 3.1). This is referred to as *epicellular* mineralization. In some cases, the individual cells become so totally encrusted in the mineral deposit that they put on enough weight to cause them to sink and form sediments. Although the deposition process is largely adventitious, the organic components of the cell wall—lipids, proteins and polysaccharides—can influence the mineralization process by acting as a general surface for precipitation. Moreover, specific sites in the cell wall are often involved because these are localized regions of metabolic efflux, or as in the case of

Table 3.1 Biologically induced mineralization in bacteria

Mechanism	Mineral	Examples
Soluble biopolymers	Mn/FeOOH	*Leptothrix*
		Pedomicrobium
Spore coats	MnOOH	*Bacillus*
Gas/ion exchange		
H_2S	Fe/CuS	*Desulfovibrio*
CO_2/pH	$CaCO_3$	*Calothrix*
pH	$MgNH_4PO_4$	*Proteus*
Membrane transport	$Ca_{10}(PO_4)_6(OH)_2$	*Streptococcus*
Enzyme activity	$(UO_2)_3(PO_4)_2$	*Citrobacter*
Electron transfer	Fe_3O_4	GS-15
	UO_2	GS-15
	Au	*Pedomicrobium*
Nucleation proteins	H_2O (ice)	*Pseudomonas*
Surface layer proteins	FeOOH	*Leptothrix*

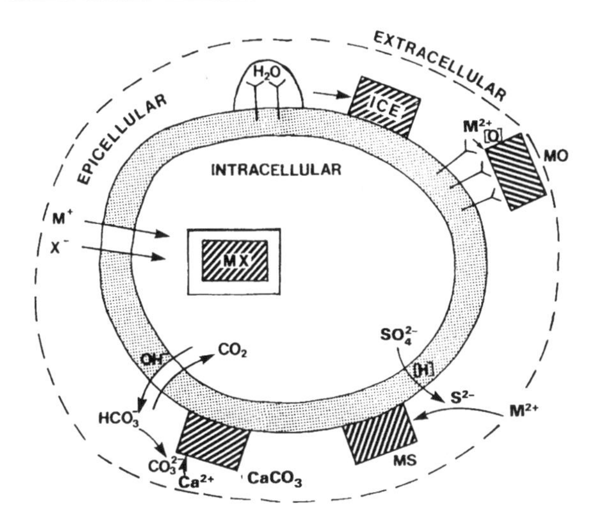

Fig. 3.1 Biologically induced mineralization reactions associated with the cell wall.

ice formation in certain bacteria, contain high concentrations of aggregated proteins that are active in nucleation.

Because the biominerals produced by biologically induced mineralization are deposited adventitiously, they are not under strict cellular control. One consequence of this is that the size, shape, structure, composition and organization of the mineral particles are often poorly defined and heterogeneous. For example, Fig. 3.2 shows irregularly shaped nanoparticles of magnetite (Fe_3O_4) produced by a bacterium referred to as GS-15. In fact, in many cases it is difficult to distinguish between the biologically induced minerals and their inorganic counterparts produced by precipitation reactions in the laboratory.

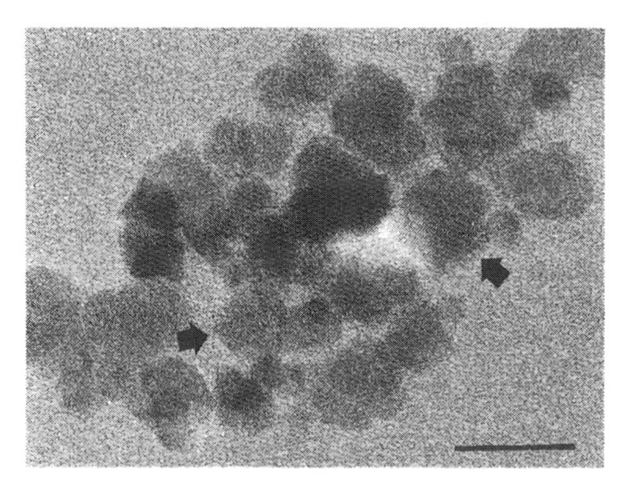

Fig. 3.2 Biologically induced magnetite nanoparticles. Arrows denote crystals showing atomic lattice planes. Scale bar, 20 nm.

Biologically induced mineralization involves the adventitious precipitation of inorganic minerals by reaction of extraneous ions with metabolic products extruded across or into the cell wall. The mineral products are closely associated with the cell wall and are crystallochemically heterogeneous.

3.2 Biologically controlled mineralization

In contrast to biologically induced mineralization, *biologically controlled mineralization* is a highly regulated process that produces materials such as bones, shells and teeth that have specific biological functions and structures, many of which are described in Chapter 2. These biominerals are distinguished by reproducible and species-specific crystallochemical properties, which include:

- uniform particle sizes
- well-defined structures and compositions
- high levels of spatial organization
- complex morphologies
- controlled aggregation and texture
- preferential crystallographic orientation
- higher-order assembly into hierarchical structures.

Biologically controlled mineralization is widespread in unicellular creatures, such as algae and protozoa, and extremely common in multicellular organisms. Only one well-documented case is known in bacteria, which is the formation of magnetite (Fe_3O_4) crystals in so-called *magnetotactic bacteria* (see also Chapter 2, Section 2.6.1). These organisms produce Fe_3O_4 nanoparticles with species-specific shapes, as shown in Fig. 3.3 (see also Fig. 7.1, Chapter 7, Section 7.1). In fact, these morphologies are so unique that they have been looked for in Martian meteorites as a possible sign of extraterrestrial life!

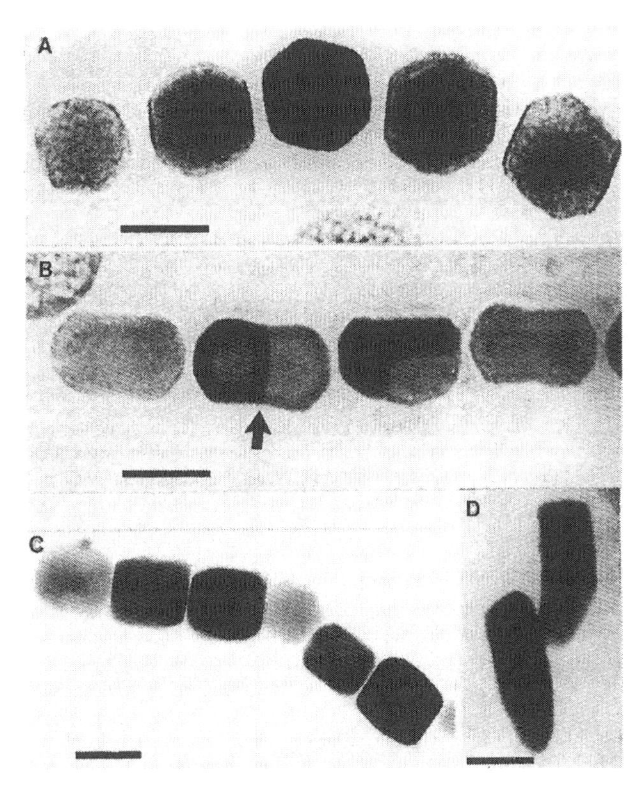

Fig. 3.3 Morphological types of bacterial magnetite single crystals: (A) cubo-octahedron; (B) elongated hexagonal prism—one crystal is twinned (arrow); (C) flat-topped hexagonal prism; (D) bullet-shaped. Scale bars, 50 nm.

Comparing the biologically controlled bacterial magnetite crystals shown in Fig. 3.3 with those formed by biologically induced mineralization (Fig. 3.2) clearly highlights the radically different consequences arising from these two fundamental processes of mineral deposition in organisms. Indeed, understanding the principles and concepts that account for biologically controlled mineralization is the central focus of this book. Moreover, the study of biologically controlled mineralization is challenging our existing notions of materials chemistry and providing novel insights. Some of these *biomineral-inspired* approaches are described in Chapter 9.

Biologically controlled mineralization involves the specialized regulation of mineral deposition and results in functional materials with species-specific crystallochemical properties.

3.3 Site-directed biomineralization

Biomineralization takes place within four main biological sites:

- *epicellular*—on the cell wall (see Section 3.1)
- *intercellular*—in the spaces between closely packed cells
- *intracellular*—inside enclosed compartments within the cell

- *extracellular*—on or within an insoluble macromolecular framework outside the cell.

Most of the examples described in this book will concern biologically controlled mineralization processes that occur in *intracellular* or *extracellular* sites. There are other examples, notably in colonies of closely packed cells, in which the minerals form on the outer surfaces of each cell and become so intermeshed that a continuous wall fills the *intercellular* spaces. This occurs for instance in coral reefs, where the colony gradually becomes entombed within a remarkable skeletal structure (Fig. 3.4).

In general, two types of assembled organic structures are used to delineate the mineralization sites associated with controlled biomineralization. Within the cells, these take the form of membrane-bounded compartments (see Section 3.3.1), whereas macromolecular frameworks (Section 3.3.2) are used in extracellular environments.

Biomineralization occurs at specific sites on, in, between or outside the cell. Intracellular and extracellular sites are often involved with biologically controlled mineralization.

3.3.1 Lipid vesicles

Many unicellular organisms deposit biominerals, such as silica and calcium carbonate, by biologically controlled processes located within intracellular *vesicles*. These specially formed microenvironments are fluid-filled sites usually delineated by a membrane containing a bilayer of *lipid* molecules with embedded proteins (see Chapter 5, Section 5.1.1, for more details).

Assembly of a vesicle membrane is based on well-known principles of physical chemistry. Amphiphilic molecules such as surfactants and lipids contain both hydrophilic and hydrophobic residues and therefore exhibit chemical schizophrenia when dissolved in a solvent. Surfactant molecules have a polar or charged headgroup and a long hydrophobic tail, so they are drawn in the shape of a tadpole or pollywog. Lipid molecules are generally more complex and have two hydrophobic tails. In water, the hydrophilic headgroups are solvated and extrovert in character, whereas the hydrophobic chains become chemical introverts intent on shielding themselves from the

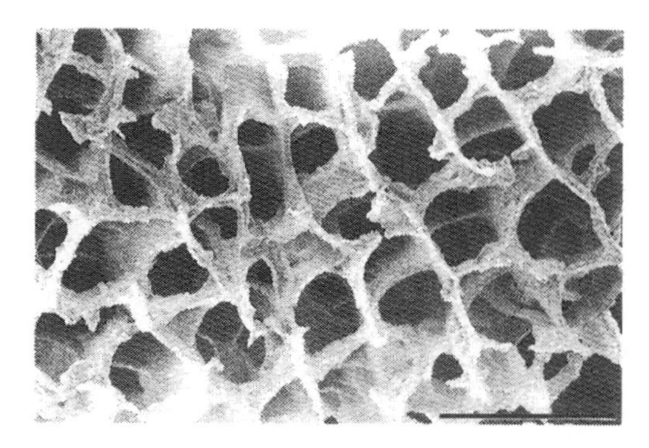

Fig. 3.4 Calcified coral. Scale bar, 500 μm.

uncomfortable polar interactions. A free energy compromise is attained by self-assembly, in which the hydrophobic residues become internalized within a supramolecular aggregate such as a micelle (Fig. 3.5). The hydrophilic domains of each molecule, on the other hand, are exposed at the surface of the micelle where they interact with the water molecules of the solvent. The overall shape of the micelle is strongly influenced by the shape of the individual molecules in the cooperative structure. When the headgroup is much wider than the alkyl chain, the molecules are wedge-shaped, and self-assemble into micellar aggregates with positive curvature as shown in Fig. 3.5. If both domains are similar in width, for example by adding a second bulky hydrophobic tail in a lipid molecule, then a planar sheet with two layers is energetically stable at high lipid concentrations (Fig. 3.6). Although the biomolecular sheets can be very long, they still have edges at which the hydrophobic tails are exposed to the water, so in the presence of large amounts of water they close in on themselves to form aqueous-filled spherical bilayer vesicles (Fig. 3.7).

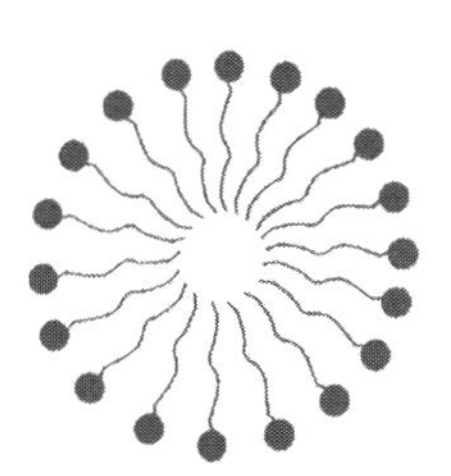

Fig. 3.5 Surfactant micelle with polar headgroups exposed.

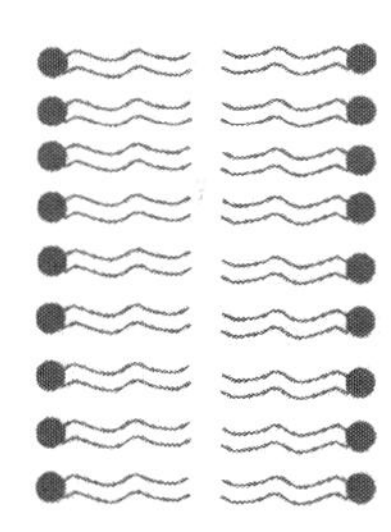

Fig. 3.6 Planar lipid bilayer (lamellar phase). The sheet is 4 to 5 nm in width.

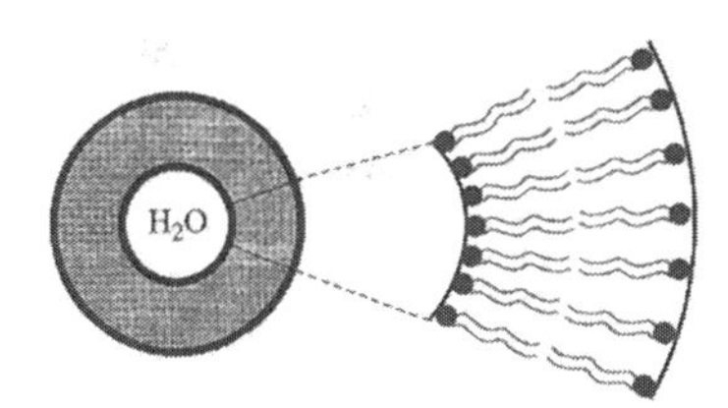

Fig. 3.7 Lipid vesicle with aqueous inner compartment and bilayer shell.

In water, groups of lipid molecules spontaneously self-assemble into bilayer vesicles. These fluid-filled spherical structures play an important role as sites for intracellular biomineralization.

3.3.2 Macromolecular frameworks

The small size of vesicles is not generally compatible with the building of large structures such as bones, shells or teeth, which are therefore constructed in extracellular spaces. These processes are precisely regulated through the activity of specialized cells that seal off a space into which an *organic matrix*, consisting of insoluble proteins and polysaccharides such as collagen or chitin, respectively, is secreted. The mineral phase is then deposited in close association with the organic matrix. For example, in the avian eggshell, calcite crystals develop at specific sites on the surface of an extracellular fibrous matrix (Fig. 3.8). In bone, hydroxyapatite crystals are located within regular gaps that occur between collagen fibres, and in seashells aragonite crystals grow between an ordered array of insoluble protein sheets. Further details of these processes are discussed in Chapter 6.

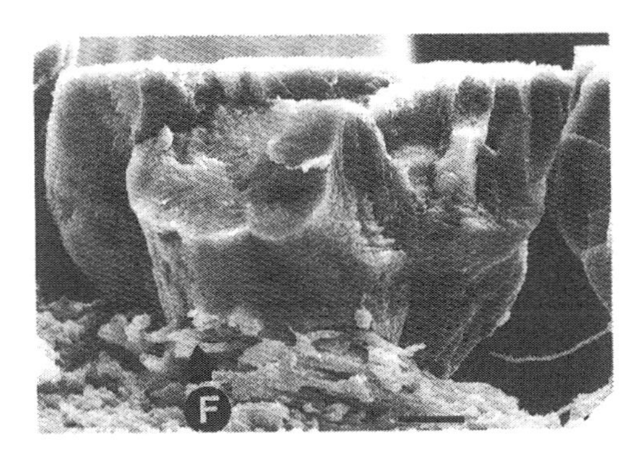

Fig. 3.8 Early stages of eggshell formation showing calcite crystal and associated macromolecular fibres (F). Scale bar, 10 μm.

Macromolecular frameworks in the form of an organic matrix are commonly used to control biomineralization in extracellular sites.

3.3.3 Site requirements

Although the mechanisms that govern the biological control of mineralization vary enormously in different systems, there are four basic requirements associated with mineralization sites such as vesicles and macromolecular frameworks. As shown in Fig. 3.9, these include:

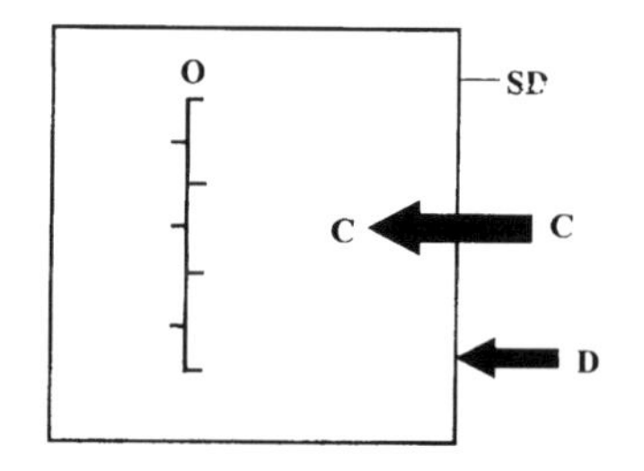

Fig. 3.9 Site-directed biomineralization involves spatial delineation (SD), diffusion-limited ion flow (D), chemical regulation (C) and organic surfaces (O).

- *spatial delineation*—for size and shape control
- *diffusion-limited ion flow*—for controlling solution composition
- *chemical regulation*—for increasing ionic concentrations
- *organic surfaces*—for controlling nucleation.

The site is sealed off from the general biological environment prior to mineralization. Spatial delineation of the intra- or extracellular space is important because it provides a physical boundary that is often used to determine the size and shape of the mineral phase. Moreover, only certain ions and molecules are allowed to diffuse through the boundary and have access to the confined space. Thus the composition of the fluid within the site is often very different from that in the surrounding medium. One important aspect of the fluid composition is that at least one ion must have a concentration raised above background levels for mineralization to occur. More specifically, the activity product of the mineral ions must be increased above the solubility product (see Chapter 4, Section 4.2). There are several chemical-based mechanisms that can elevate the ionic concentrations, and in general they rely on the *pumping* of ions, such as Ca^{2+}, into the confined space (see Chapter 5, Section 5.3, for details). Clearly, by controlling the rates of ionic diffusion, for example through a vesicle, cell membrane or semipermeable organic framework, the solution chemistry leading to mineralization can be fine-tuned. However, even greater levels of regulation may be required if the initial stages of mineral deposition (*nucleation*) are to be precisely controlled. This is achieved by incorporating an organic surface into the site that is capable of interacting with metal ions and orchestrating their transformation into the mineral phase (see Chapter 6, Section 6.7).

Site-directed biomineralization involves a preformed spatially enclosed environment, which acts as a physical boundary and a diffusion-limited zone that is capable of maintaining a particular ionic composition. The site also contains mechanisms for raising the concentrations of certain ions and an organic surface for controlling nucleation.

3.4 Control mechanisms

At a fundamental level, the processes of controlled biomineralization reflect many of the general features that distinguish biology. They are governed by the gene pool, driven by bioenergetic processes and adapted to environmental influences, so there are many levels of regulation embedded into a complex

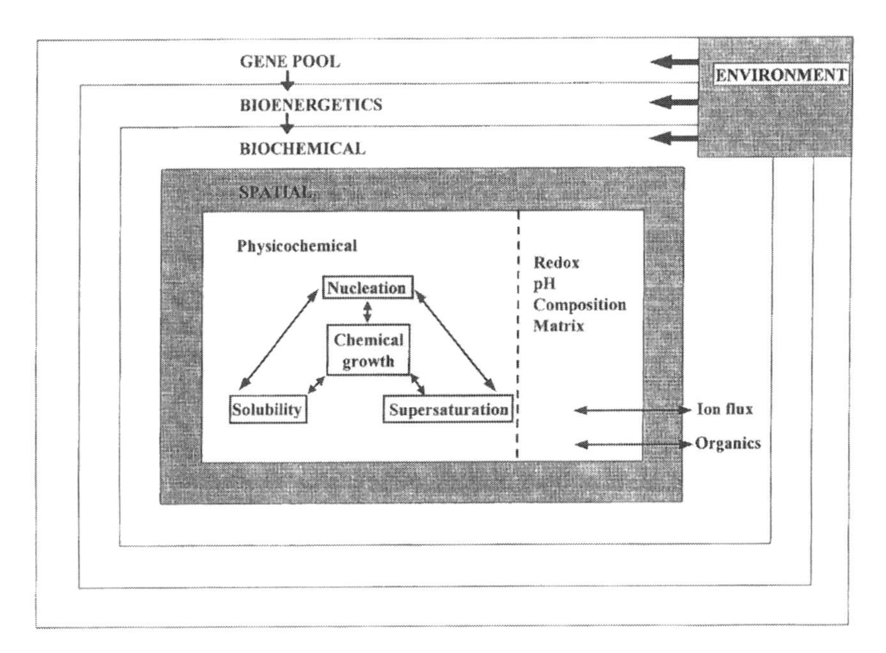

Fig. 3.10 Levels of regulation in biologically controlled mineralization.

interactive network (Fig. 3.10). Very little is known about how these generic processes operate. Our focus is more specific and accessible, and concerns five key control mechanisms, namely:

- chemical
- spatial
- structural
- morphological
- constructional.

These are briefly discussed in turn in the following sections of this chapter, and then developed in detail in Chapters 4 to 8.

Control mechanisms in biomineralization involve the regulation of chemistry, space, structure, morphology and construction.

3.4.1 Chemical control

Chemistry is fundamental to the start, stop and go of controlled biomineralization. In the case of crystalline biominerals, four fundamental physicochemical factors—*solubility*, *supersaturation*, *nucleation* and *crystal growth*—are involved (see Chapter 4 for details). In general, the solubility of the inorganic mineral is a crucial factor in determining the thermodynamic conditions for precipitation, and the extent to which a solution is out of equilibrium is given by the supersaturation, which in turn influences the rates of nucleation and growth. Each of these factors can be controlled at the molecular level by regulating the chemical conditions of the biological solutions contained within the mineralization sites. This is principally achieved in biological systems by the regulation of *ion transport*, and by specialized molecules that act as *promotors* or *inhibitors* of crystal growth and phase transformation.

Solubility, supersaturation, nucleation and crystal growth are chemically controlled in biomineralization by coordinated ion transport and molecular-based inhibitors and promotors.

3.4.2 Spatial control

Spatial control refers principally to the regulation of the size and shape of biominerals by confining the process to specific boundary spaces such as vesicles or a porous organic framework (see Chapter 5). These organic structures are assembled before mineralization through covalent and non-covalent interactions between the molecules. This process is referred to as *supramolecular preorganization*. However, as we described in Section 3.3.3, spatially delineated sites are not just passive enclosures but active in moving selective ions and molecules into the mineralization zone. Thus, spatial control is often related to both the physical dimensions of the site and mechanisms for chemical control.

The control of space in biomineralization occurs through the supramolecular preorganization of organic molecules, and impacts on the size and shape of mineral deposits and the chemical mechanisms of their deposition.

3.4.3 Structural control

One of the most fascinating aspects of biologically controlled mineralization is the ability of certain systems to produce structures in which the inorganic crystals are not only associated specifically with an organic matrix but also aligned preferentially with respect to the macromolecular surface. In many cases, nucleation on the organic surface results in the alignment of one specific axis of the unit cell perpendicular to the surface of the matrix sheets or fibres. The alignment is often less precise in directions parallel to the matrix surface so the crystals are rotated around a constant crystallographic axis to produce a mosaic of partially oriented particles (Fig. 3.11). Occasionally, two crystal axes are specifically aligned with respect to the

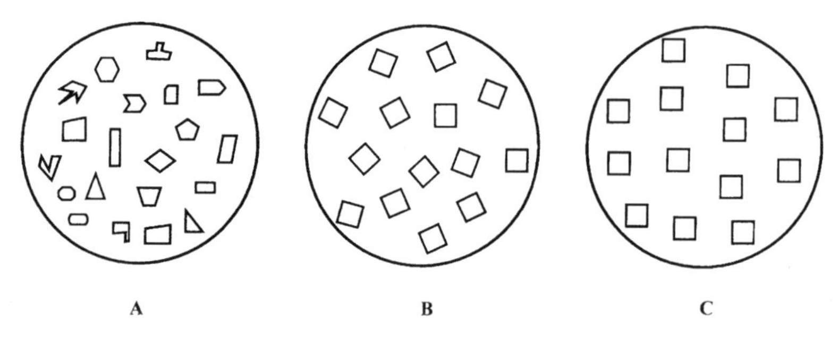

Fig. 3.11 Nucleation of biominerals on organic surfaces: (A) non-oriented; (B) mosaic with crystals aligned only perpendicular to the matrix; (C) iso-oriented array with 3-D crystallographic alignment.

underlying molecular structure of the matrix. As shown in Fig. 3.11, this produces an iso-oriented array of spatially separated crystals that have the same crystallographic orientation in three dimensions.

These observations indicate that there is precise structural control over mineral deposition by organic surfaces located within the mineralization sites. The general hypothesis is that the macromolecular surface contains appropriate chemical groups that bind or concentrate metal ions in structural arrangements that resemble the ordering of the same ions in a particular crystal face. In this way, the matrix acts as an *organic template* for inorganic nucleation. The fundamental principle governing this process is *interfacial molecular recognition*, which is described in detail in Chapter 6, Section 6.7.1.

Structural control in biomineralization involves the preferential nucleation of a specific crystal face or axis by molecular recognition at the surface of an organic matrix.

3.4.4 Morphological control

Many of the biominerals described in Chapter 2 have complex morphologies that bear little or no resemblance to the same minerals formed in chemical and geological systems. For example, amorphous minerals such as silica usually precipitate in inorganic systems as formless gels or spherical colloidal particles, and not elaborate arrangements such as the lace-like scales and shells of unicellular organisms shown in Fig. 3.12. This sculpturing of inorganic form is even more remarkable when we consider crystalline minerals because the biomineral shapes challenge the conventional view of morphology as the macroscopic manifestation of the geometric arrangement of atoms in the unit cell. It is almost inconceivable that microscopic architectures like the trumpet-shaped calcium carbonate coccoliths shown in Fig. 2.3 of Chapter 2 can arise from the regular geometric-based symmetry of inorganic crystals.

As a general principle, we can consider the complex shapes of biominerals to originate from the *vectorial regulation* of crystal growth and precipitation within or between organic structures such as vesicles and polymeric frameworks. That is, the mineral growth process is controlled by organic bound-

Fig. 3.12 Silica scales produced in *silicoflagellates*.

aries that change in size and shape with time so that the inorganic phase is progressively routed along specific directions set by a biological programme rather than determined by the intrinsic crystallographic axes of the unit cell. The running of this *patterning* programme is complex and dependent on many interrelated processes occurring in and outside the mineralization site. The origin and development of biomineral form is referred to under the general term *morphogenesis*, and is discussed in detail in Chapter 7.

Control over the complex form of biominerals is achieved by patterning processes (morphogenesis) that give rise to time-dependent vectorial growth.

3.4.5 Constructional control

The incredible complexity of biomineralized tissues such as bones and shells is associated with the controlled construction of hierarchical architectures that involve the assembly of mineral-based building blocks into a series of progressively higher-order structures. In bone, the primary building units are collagen fibres and their associated tiny crystals of hydroxyapatite. These are woven along with bone cells into a number of structural tapestries that exhibit order on longer length scales. In turn, these new structures are used as modules for further construction until eventually they become embedded by the hierarchical process. The integration of these constructional processes across multiple length scales is referred to as *biomineral tectonics*, and further details are described in Chapter 8.

Higher-order structures in biomineralization are constructed by a series of integrated processes that extend across a range of length scales.

3.5 General model

The control processes summarized in Section 3.4 can be incorporated into a general model of biomineralization, which is illustrated in Fig. 3.13. Mineralization occurs in localized spaces, such as within the microcompartments of intracellular vesicles or fluid-filled voids of an extracellular organic matrix. The dimensions and shapes of these compartments influence the size and initial morphology of the mineral particles. In both cases, supersaturated solutions have to be attained to initiate nucleation.

In extracellular mineralization (Scheme A in Fig. 3.13), the polymeric framework is secreted away from the neighbouring cells so the supersaturation level of the fluid is established and maintained by the transport of ions over relatively large distances. This is achieved by the passive diffusion of hydrated ions pumped out of the adjacent cells, or through the use of vesicles loaded with high concentrations of certain ions. These ions become selectively encapsulated in the vesicle by membrane transport processes that often pump ions against gradients established across the lipid membrane. For example, in Scheme A, this mechanism produces vesicles containing increased concentrations of the cation M^{n+}, whereas the corresponding

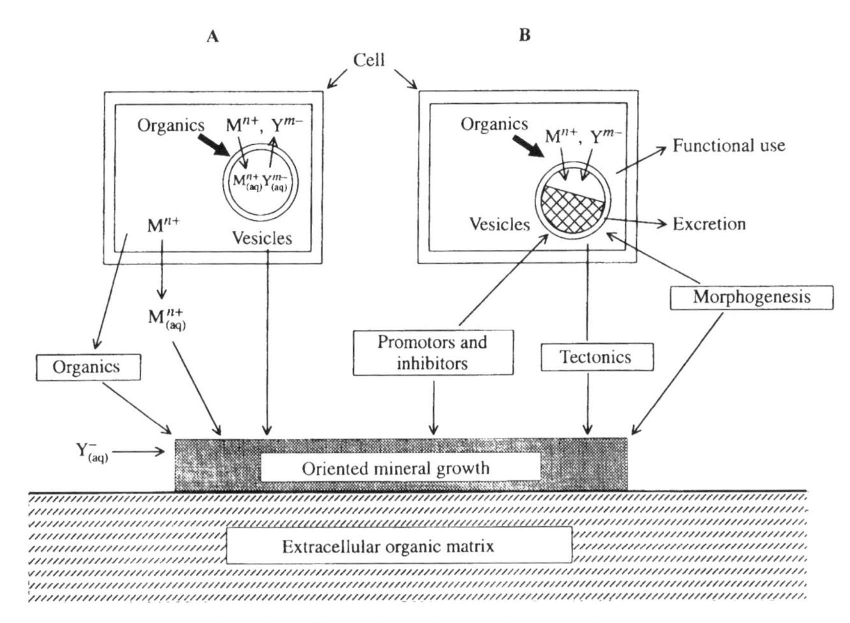

Fig. 3.13 General model of biomineralization. See text for details.

mineral anion, Y^{m-}, is excluded either by active outward transport or slow diffusion kinetics. The two ions are only allowed to combine to produce the mineral phase after the vesicles reach the organic matrix, where they are destroyed or disrupted by promoter molecules. Alternatively, in intracellular mineralization (Scheme B in Fig. 3.13), both M^{n+} and Y^{m-} ions are transported into the vesicle so that precipitation of the mineral occurs directly within the microcompartment.

As well as the spatial and chemical controls that are intrinsic to the vesicle and matrix sites, the lipid membrane and macromolecular surface, respectively, can influence the nucleation stage of biomineralization. In certain cases, specific interfacial interactions result in structural matching between the inorganic nuclei and functional groups on the organic surface such that the inorganic phase is aligned along preferred crystallographic directions. Subsequent growth of the mineral is chemically controlled by auxiliary molecules residing inside the vesicles or in the extracellular fluid. These control the growth rate by inhibition or promote phase transformations, such as the conversion of amorphous calcium phosphate to crystalline hydroxyapatite. In addition, the mineral deposits become shaped into complex forms by patterning mechanisms arising from morphogenetic processes acting on the vesicles or extracellular matrix.

Intracellular and extracellular mineralization can be coupled together for the construction of higher-order assemblies. Usually, the intravesicular deposits remain functionally active within the cell but in certain circumstances can be ejected through the cell membrane not for excretion but for use as prefabricated building units. In this case, the mineral-containing vesicles are transported to a remote extracellular site where the inorganic deposits are released and spatially organized in association with an organic matrix (Fig. 3.14).

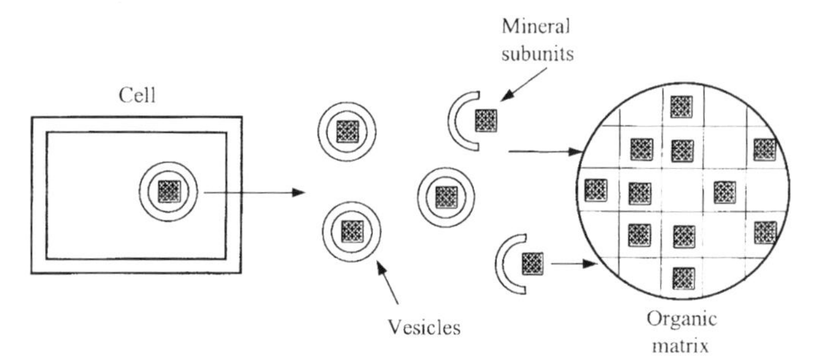

Fig. 3.14 Higher-order assembly in biomineralization.

A general model of biomineralization involves both vesicles and organic matrices as potential intra- and extracellular mineralization sites, respectively, that are under chemical, spatial, structural and morphological control. Vesicles are often involved in the long-range transport of ions or mineral deposits to the extracellular matrix. The latter can be used as prefabricated building blocks for the construction of higher-order mineral assemblies.

3.6 Summary

In this chapter we have defined some general principles involved with the control of biomineralization. Although these mechanisms do not generally act independently of each other, they can be associated with particular types of biomineralization processes. Except for the principle of chemical control, which is a pivotal aspect of all processes of controlled precipitation (crystallization) in biological milieus, we can associate the mechanisms described above under spatial, structural, morphological and constructional control with four principal processes of biomineralization (Table 3.2). These are, respectively:

- boundary-organized biomineralization
- organic matrix-mediated biomineralization
- morphogenesis
- biomineral tectonics.

Because each of these processes is underpinned by the chemical control of biomineralization, the next chapter describes the principal concepts of controlled crystallization and their applications in biomineralization. A detailed discussion of each of the four biomineralization processes listed above then follows in Chapters 5 to 8.

Along with the general process of chemically controlled precipitation, there are four specific biomineralization processes—boundary-organized biomineralization, organic matrix-mediated biomineralization, morphogenesis and biomineral tectonics—that regulate, respectively, the spatial, structural, morphological and constructional aspects of mineral deposition in biological systems.

Table 3.2 The main types of biomineralization processes

Process	Control mechanism	Concepts	Properties	Chapter
Precipitation (crystallization)	Chemical	Solubility Supersaturation Nucleation Growth	Solution composition Promotion Inhibition Phase transformation	4
Boundary-organized biomineralization	Spatial	Supramolecular preorganization	Physical boundary Diffusion-limited site Ion transport Size and shape Organization	5
Organic matrix-mediated biomineralization	Structural	Interfacial molecular recognition	Site-directed nucleation Oriented nucleation Supporting framework Mechanical design	6
Morphogenesis	Morphological	Vectorial regulation	Complex form Time-dependent form Patterning	7
Biomineral tectonics	Constructional	Multilevel processing	Higher-order assembly Hierarchical structures Integrative building modules Adaptive structures and functions	8

Further reading

Addadi, L. and Weiner, S. (1992). Control and design principles in biological mineralization. *Angew. Chem., Int. Ed. Engl.*, **31**, 153–169.

Krampitz, G. and Graser G. (1988). Molecular mechanisms of biomineralization in the formation of calcified shells. *Angew. Chem., Int. Ed. Engl.*, **27**, 1145–1156.

Lowenstam, H. A. (1981). Minerals formed by organisms. *Science*, **211**, 1126–1131.

Lowenstam, H. A. and Weiner, S. (1989). *On biomineralization*, pp. 25–49. Oxford University Press, New York.

Mann, S. (1988). Molecular recognition in biomineralization. *Nature*, **332**, 119–124.

Simkiss, K. (1986). The processes of biomineralization in lower plants and animals—an overview. In *Biomineralization in lower plants and animals* (ed. Leadbeater, B. S. C. and Riding, R.), pp. 19–37. Systematics Association Vol. 30. Oxford University Press, Oxford.

Watabe, N. (1981). Crystal growth of calcium carbonate in the invertebrates. *Prog. Cryst. Growth Charact.*, **4**, 99–147.

4 Chemical control of biomineralization

The high level of regulation associated with biologically controlled mineralization is fundamentally dependent on the chemical control of inorganic precipitation and crystallization. In this chapter, we review the key principles of precipitation—usually from the standpoint of 'crystallization'—and illustrate their relevance to biomineralization. The aim is to introduce the fundamental chemical concepts that are pivotal to the specific biological control mechanisms described in Chapters 5 to 8.

4.1 Solubility

The *solubility* of an inorganic salt is the number of moles of the pure solid that will dissolve in a litre of solvent at a given temperature. Dissolution occurs when the free energy required to disrupt the lattice bonding, ΔG_L, is offset by the free energy released in the formation of aqueous species, such as hydrated ions (ΔG_H), ion pairs (ΔG_{IP}) and complexes (ΔG_C). The free energy of solution, ΔG_S, is given by:

$$\Delta G_S = \Delta G_L - (\Delta G_H + \Delta G_{IP} + \Delta G_C)$$

There are a number of empirical equations that predict the value of ΔG_S on the basis of the size and charge of the cation and anion (see further reading). In general, an ionic compound will be highly insoluble for cation and anion radii of similar dimension. This is because the matching in size results in a closely packed structure that maximizes the electrostatic interactions and hence increases the lattice energy.

For many ionic salts in pure solution, ΔG_L and ΔG_H are large energy terms that have magnitudes dependent on both enthalpic and entropic factors. They are also strongly influenced by changes in the structure and composition of the mineral phase deposited. For example, ions such as Na^+, NH_4^+, K^+, Mg^{2+}, Fe^{2+}, CO_3^{2-} and F^-, in particular, are readily incorporated into the lattice structure of hydroxyapatite (HAP), where they have a pronounced effect on the solubility. This is the principal reason why fluoride has been so successful in preventing tooth decay. If the HAP crystals of enamel are attacked by weak acids (rhubarb or certain soft drinks would do the trick), then in the presence of fluoride the dissolved calcium and phosphate ions immediately reprecipitate as the less soluble fluoroapatite (FAP) phase. Because both the saliva and plaque fluids are always undersaturated with respect to FAP (Fig. 4.1), substitution of some of the OH^- ions in the HAP lattice for F^- and the resulting changes in mineral solubility are sufficient to inhibit enamel demineralization and reduce dental caries.

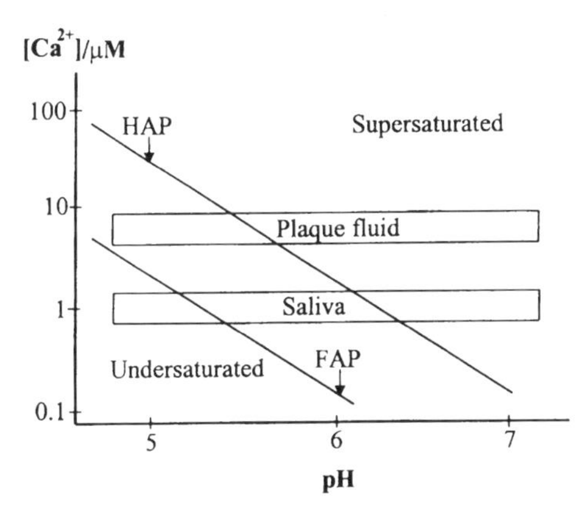

Fig. 4.1 Plot of solubility against pH, measured in terms of $[Ca^{2+}]$ concentration, for HAP and FAP minerals. The actual $[Ca^{2+}]$ levels for saliva and plaque fluids are also shown.

The solubility of an inorganic salt depends on the balance between lattice energy and ion solvation and complexation in aqueous solution. Solubility is a key factor in the biological demineralization of calcium phosphates such as hydroxyapatite.

4.2 Solubility product

The solubility (s) of an inorganic salt can be related to an equilibrium constant, the *solubility product*, K_{sp}, provided that there is negligible complexation. For the general case of an ionic solid containing univalent ions,

$$M_nX_m(\text{solid}) \rightleftharpoons nM^+_{(aq)} + mX^-_{(aq)}$$

and

$$K_{sp} = \{M^+\}^n \cdot \{X^-\}^m$$

where $\{M^+\}$ and $\{X^-\}$ are the effective concentrations (*activities*) of ions in a solution in equilibrium with the solid phase. The solubility product therefore is a measure of the *activity product* at equilibrium and represents the baseline for determining the thermodynamic condition for inorganic precipitation. Clearly, if the solubility product is less than the activity product (AP) of a solution then precipitation will occur until K_{sp} is equal to AP.

The solubility product can be determined by measuring the solubility because

$$\{M^+\} = n \cdot s, \{X^-\} = m \cdot s$$

Moreover, values of K_{sp}, such as those listed in Table 4.1, can be used to calculate the free energy of solution ΔG_S, by

$$\Delta G_S = -RT \ln K_{sp}$$

where R is the gas constant and T the temperature.

Table 4.1 Solubility products for biominerals

Mineral	Solubility product ($\log K_{sp}$)	
Calcium carbonate		
Monohydrite	−7.39	
Vaterite	−7.60	
Aragonite	−8.22	
Calcite	−8.42	
Calcium phosphate		
Brushite ($CaHPO_4 \cdot 2H_2O$)	−6.4	−3.2*
Octacalcium phosphate ($Ca_8H_2(PO_4)_6$)	−46.9	−6.7*
Hydroxyapatite ($Ca_{10}(PO_4)_6(OH)_2$)	−114.0	−7.1*
Fluoroapatite ($Ca_{10}(PO_4)_6F_2$)	−118.0	−7.4*
Amorphous silica		−2.7*
(Quartz)		−3.7*
Iron oxides		
Ferrihydrite	−37.0	
Goethite (α-FeOOH)	−44.0	
(Hematite, α-Fe_2O_3)	−42.5	
Group 2A sulfates		
Gypsum ($CaSO_4 \cdot 2H_2O$)	−5.03	
Barite ($BaSO_4$)	−9.96	
Celestite ($SrSO_4$)	−7.40	

*Values quoted as mol l^{-1}.

The relationship between solubility and solubility product is not so clear cut when applied to the deposition of biominerals. For example:

- The solubility product criterion is difficult to apply meaningfully when there is substantial complex formation, as for example in the dissolution of biogenic iron oxides which produces various types of oxo- and hydroxo-complexes.
- The concept of solubility product cannot be applied to covalent solids precipitated from neutral species. Thus, it is not applicable in silica biomineralization which occurs predominantly by the polycondensation of silicic acid, $Si(OH)_4$. In this case, only the solubility can be measured.
- The determination of K_{sp} is exceedingly difficult for biological fluids because the measured ionic concentrations often deviate markedly from the ionic activities due to extensive complex formation with a wide range of organic molecules that are present in the solutions. Moreover, the concentrations of these complexing agents often change with time.
- The solubility is not a constant but increases with diminishing crystal size because small crystals have high surface-to-volume ratios and therefore the surface energy begins to dominate over the lattice energy. Thus, small crystals in the presence of large crystals preferentially dissolve whereas the large crystals continue to grow—this process of redistribution is usually called *Ostwald ripening*.

- The solubility product is a thermodynamic concept and takes no account of kinetic effects. Many biominerals are kinetically stabilized from dissolution by being wrapped up in insoluble organic sheaths.

The solubility product (K_{sp}) is a critical factor in determining the thermodynamic limit for the onset of inorganic precipitation. When the solubility product is less than the activity product (AP) of a solution then precipitation will occur until K_{sp} = AP.

4.3 Supersaturation

The difference between the activity product (AP) and the equilibrium position as defined by K_{sp} is an indicator of the level of *supersaturation* of a solution. The relative supersaturation, S_R, is defined as

$$S_R = AP/K_{sp}$$

and the absolute supersaturation, S_A, is given by

$$S_A = (AP - K_{sp})/K_{sp}$$

The difference in chemical potential ($\Delta\mu$, where μ is the change in Gibbs free energy per unit change in composition) between the supersaturated solution and a solution at equilibrium with the solid is related to S_R by

$$\Delta\mu = kT \ln S_R$$

where k is the Boltzmann constant and T is temperature. Clearly, as S_R increases, the thermodynamic driving force for precipitation, $\Delta\mu$, rapidly increases because it is dependent on $\ln S_R$.

In inorganic systems, supersaturation can be achieved by several means, for example through chemical reactions, temperature changes, or variations in composition associated with evaporation of the solvent, changes in ionic activities, etc. As we discussed in Chapter 3, Section 3.3, biological systems seal off diffusion-limited spaces so that the transport of ions and their chemical activities can be fine-tuned to achieve appropriate levels of supersaturation for biomineralization to occur. More details about this process can be found in Chapter 5, Section 5.2.

Supersaturation is a measure of to what extent a solution is out of equilibrium and represents the thermodynamic driving force for inorganic precipitation. Supersaturation is highly regulated in biology through the process of boundary-organized biomineralization.

4.4 Nucleation

The thermodynamic driving force for inorganic precipitation, supersaturation, is often offset by the kinetic constraints of nucleation. *Homogeneous nucleation* occurs due to the spontaneous formation of nuclei in the bulk of the supersaturated solution, whereas *heterogeneous nucleation* involves the formation of nuclei on the surfaces of a substrate present in the aqueous medium. Despite the fact that homogeneous nucleation does not generally occur in biomineralization, the principles of this process form the necessary background for the understanding of nucleation at the surface of an organic matrix.

Every synthetic chemist knows the frustration of making a new compound but not being able to crystallize it even though the solution is supersaturated. Although molecular aggregates are continually forming in the supersaturated solution they are also falling apart at a rate that prevents the growth of the solid phase. This is because the aggregates grow against a gradient of free energy that is required to create the new solid–liquid interface. Only when the expenditure of this interfacial energy (ΔG_I) is balanced by the energy released in the formation of bonds in the bulk of the aggregate (ΔG_B) is a stable nucleus attained. The clusters then continue to grow because the overall change in free energy becomes progressively more negative as the aggregates increase further in size.

The free energy of formation of a nucleus, ΔG_N, is therefore given by the difference between the surface (interfacial) and bulk energies:

$$\Delta G_N = \Delta G_I - \Delta G_B$$

The interfacial energy is always positive and dependent on the surface area, whereas the bulk energy is negative and a function of volume. For the classical case of a spherical nucleus,

$$\Delta G_I = 4\pi r^2 \sigma$$

where σ is the interfacial free energy per unit surface area, and

$$\Delta G_B = 4\pi r^3 \Delta G_v / 3V_m$$

where ΔG_v represents the free energy per mole associated with the solid–liquid phase change, and V_m is the molar volume.

The main point to note is that plots of ΔG_I and ΔG_B as functions of r, the size of the cluster, are very different—not only do they have positive and negative values, but they also vary with r^2 and r^3, respectively. This means that the combination of the two functions, which equates to ΔG_N, goes through a free energy maximum (ΔG_N*) at a critical cluster size (r*), as shown in Fig. 4.2. This corresponds to the *activation energy for homogeneous nucleation*, which for a spherical nucleus can be calculated as

$$\Delta G_N^* = 16\pi\sigma^3\nu^2/3(kT \ln S_R)^2 \qquad r^* = 2\sigma V_m/\Delta G_v$$

where ν is the molecular volume (see further reading).

The rate of homogeneous nucleation, J_N, is given by the classical equation with a pre-exponential factor, A,

$$J_N = A \exp(-\Delta G_N^*/kT)$$

Measurable nucleation rates—defined as the number of nuclei formed in a unit volume per second—can be very high, around 10^6 to 10^9 m^{-3} s^{-1}, with values of A of the order of 10^{36} m^{-3} s^{-1}.

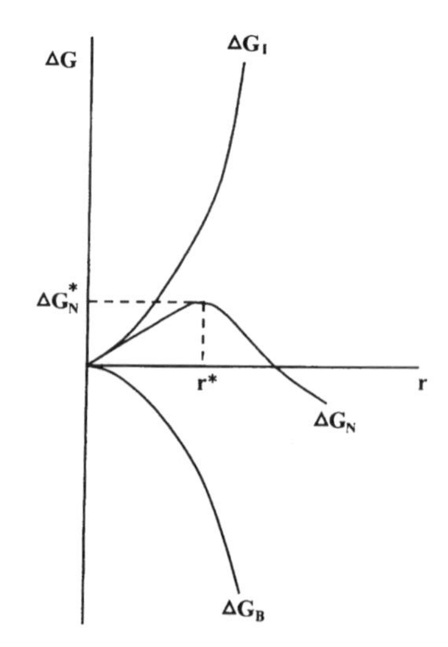

Fig. 4.2 Free energy of nucleation (ΔG_N) as a function of cluster size (r).

The above equations are important because they indicate which factors need to be influenced by biological systems if nucleation is to be regulated in biomineralization. These include the following.

- Increases in supersaturation (S_R) decrease the activation energy for nucleation (ΔG_N*) and increase the rate of nucleation (J_N). As ΔG_N* is a function of $(\ln S_R)^{-2}$, in pure solution the increase in nucleation rate can be catastrophic with a sudden runaway in the crystallization process at a certain value of supersaturation (S_R*) (Fig. 4.3). Biological systems therefore need to fine-tune the level of supersaturation within a relatively narrow window of ionic concentrations to maintain control over mineralization.
- ΔG_N* is proportional to the cube of the interfacial energy (σ^3) of the transient molecular aggregate, so relatively small changes in σ can have a marked effect on nucleation rates.
- The critical nucleus size (r*) is reduced for lower values of σ, assuming that the free energy per mole associated with the solid–liquid phase change (ΔG_v) is constant for clusters of different size.

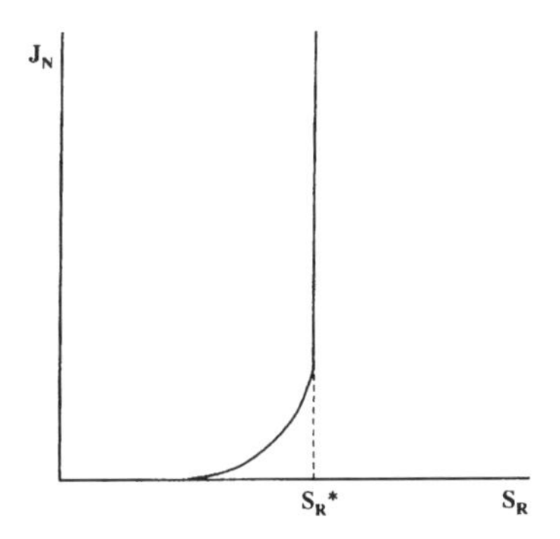

Fig. 4.3 Rate of nucleation (J_N) plotted against relative supersaturation (S_R).

The strong dependence of nucleation and critical nucleus size on interfacial energy (σ) is the reason why heterogeneous nucleation is the norm in most situations. The presence of an external substrate—a speck of dust or hair in the crystallization dish, for example—can significantly reduce σ and hence increase the rate of nucleation at a given level of supersaturation. This is why synthetic chemists like to scratch the glass walls of their reaction containers, as well as their chins, when trying to crystallize a new compound. Heterogeneous nucleation occurs, therefore, at lower supersaturation levels than those required for homogeneous nucleation because the nuclei are stabilized by attachment to the foreign surface, particularly if there is *chemical and structural complementarity*. If all the extraneous particles have equal nucleation efficiencies, then, as shown in Fig. 4.4, the rate of nucleation (J_N) can be controlled at a constant level over a large range of supersaturations (S_R) provided that the threshold for homogeneous nucleation (S_R*) is not breached. Such a process is fine-tuned in biomineralization by using organic surfaces to regulate the activation energy for nucleation through specific interfacial interactions (see Chapter 6, Section 6.7, for further details).

The activation energy and rate of nucleation are determined by the interfacial energy of the critical nucleus and the level of supersaturation. These factors can be biologically controlled in biomineralization through the evolutionary design of organic matrices and the membrane regulation of ion concentration gradients.

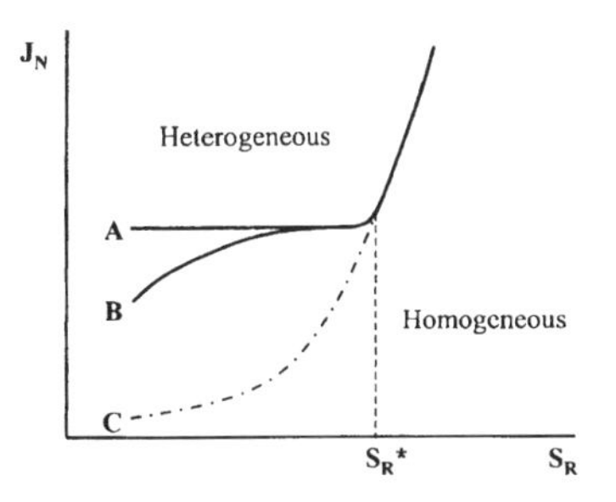

Fig. 4.4 Rate of nucleation (J_N) as a function of relative supersaturation (S_R) in the presence of extraneous particles with (A) equal, and (B) variable nucleation efficiencies; and (C) without extraneous particles.

4.5 Oriented nucleation—epitaxy

The oriented overgrowth of inorganic crystals on insoluble substrates is often observed in geological deposits such as rocks and minerals. Even though the substrate consists of a mineral with different structure and chemical composi-

Table 4.2 Epitaxial deposition of inorganic crystals on inorganic substrates. A positive percentage misfit value indicates that the overgrowth lattice is larger than the substrate lattice

Substrate	Overgrowth	Lattice misfit %
PbS	NaI	8
	KCl	5
	NaBr	−1
	NaCl	−6
	AgBr	−4
	AgCl	−7
$CaCO_3$	RbBr	7
	RbCl	3
	KBr	3
	NaI	1
	KCl	−2
	NaBr	−7
CaF_2	NaBr	8
	NaCl	3
	LiBr	0
	LiCl	−6
NaCl	NaBr	6
	NaCN	6
	AgBr	3
	AgCN	3
	AgCl	1
	KF	−5

tion to that of the overgrowth, the two phases are crystallographically oriented with respect to each other. This phenomenon, usually referred to as *epitaxy*, arises generally from a high degree of lattice matching between the overgrowth and substrate (Table 4.2).

Epitaxial deposits can be divided into two groups, those in which the orientation is established directly by nucleation, and those in which the alignment is governed by rearrangements during the growth of the initially deposited nuclei. In both cases, the substrate exerts control over the crystallographic orientation of the overgrowth by the degree of lattice matching. In nucleation-controlled epitaxis, the alignment and structure of the nuclei are determined by the free energy of the substrate–overgrowth interface, and how this changes for different orientations of the two lattices. The interfacial energy will be at a minimum when there is a precise atomistic matching of the two structures, and increasing the misorientation of the two lattices will result in a rise in the free energy. If the interfacial energy (σ) increases rapidly with misorientation (φ) then a distinct free energy minimum arises for a specific crystallographic alignment and oriented nucleation takes place. This is shown by curve A in Fig. 4.5, while the interfacial energies represented by curves B and C are less sensitive to changes in φ.

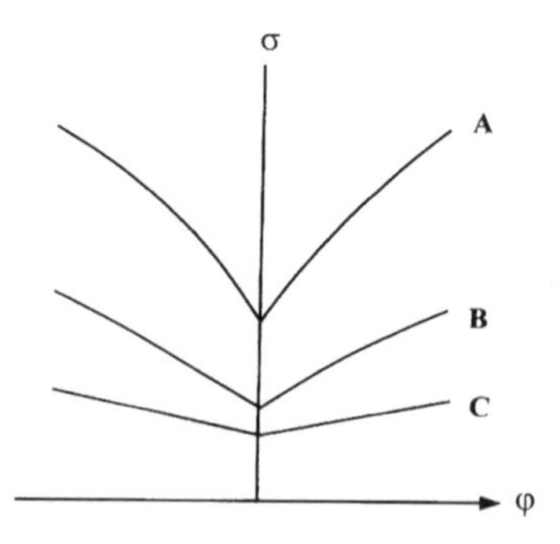

Fig. 4.5 Plot of interfacial energy (σ) as a function of lattice mismatch (φ) in epitaxy.

In some systems, however, a high degree of lattice mismatching occurs, with values outside the tolerance limits usually described for epitaxis (Table 4.3). One possibility is that the surface layers of the overgrowth are flexible enough to be distorted to fit the substrate lattice with an approximate one-to-one correspondence. It is more likely, however, that the initially formed layers of the overgrowth contain a large number of structural defects that accommodate the strain energy associated with the lattice misfit between the bulk crystal and the stretched surface. When this strain energy gets too high and the defects start to propagate into the bulk of the overgrowth, then a dramatic change from the initial orientation can occur by slipping and sliding of the bulk lattice. For example, Ag films grown on NaCl crystals have a thin interfacial region that corresponds to a 10 per cent lattice mismatch, whereas the bulk of the overgrowth has a misfit of nearly 30 per cent.

This discussion is relevant to biomineralization because many biominerals are oriented with respect to the surface of an underlying organic matrix as we mentioned in Chapter 3, Section 3.4.3. Some of the key questions therefore concern the role of epitaxy in biomineralization and whether atomistic ordering processes can take place on structured organic surfaces, and if so, how? Chapter 6, Section 6.7, looks at this important problem in detail.

Epitaxis refers to the oriented overgrowth of a crystalline phase on an insoluble crystalline substrate and is associated with the concept of lattice matching. It provides structural control of the nucleation process and is a central concept in organic matrix-mediated biomineralization.

4.6 Crystal growth

The growth of an inorganic crystal from pure solution requires the continual addition of ions to the surface and their subsequent incorporation into lattice sites. Much of the theory of crystal growth goes back to the beginning of the twentieth century and is therefore well documented in numerous textbooks

Table 4.3 Oriented overgrowth of inorganic crystals on substrates with large lattice misfits

Substrate	Overgrowth	Lattice misfit %
NaCl	NaI	15
	KCN	16
	NH_4Cl	16
	KBr	17
	NH_4Br	23
	KI	25
	NH_4I	29
	RbI	30
PbS	RbBr	15
	KI	18
CaF_2	KBr	21

(see further reading). However, the subject has received a new lease of life in recent years with the advent of atomic force microscopy, which can be used to image crystal surfaces in direct contact with a supersaturated solution. It is now possible to visualize the collective atomic motions and manoeuvres, and elucidate the underlying mechanisms of crystal growth. Movies of crystal growth may not yet have reached the video shops but they remain fascinating to watch with or without a cold beer.

Termination of growth generally occurs when the supersaturation level falls to the equilibrium level defined by the solubility product. Occasionally, crystals stop growing due to the accumulation of large numbers of surface defects that cannot be assimilated into the bulk structure by surface relaxation. Another possibility is that the crystal surface becomes blocked by the overgrowth of a more soluble secondary phase, for example a polymorph or hydrated structure that nucleates as the conditions change in the supersaturated solution. In this case, although the solution remains supersaturated with respect to the first mineral precipitated, growth no longer occurs because the surface phase is at equilibrium.

Crystal growth and termination are dependent on the level of supersaturation and occur through surface-controlled processes.

4.6.1 Mechanisms

In general, most theories of growth concern large crystals growing in pure solution with stable bulk structures, and usually express the rate of growth, J_G, in the form of a power law,

$$J_G = k(S_A)^x$$

where k = rate constant, S_A = absolute supersaturation raised to the power x, the value of which is dependent on the mechanism of the rate-determining step. There are four main scenarios:

- mass transport and diffusion-limited growth at very high values of supersaturation ($x = 1$)
- polynucleation of surface growth islands at high supersaturation ($x > 2$)
- layer-by-layer growth at moderate supersaturations ($x = 1$)
- screw dislocation growth at low supersaturation ($x = 2$).

For most purposes, crystal growth is considered as a *secondary nucleation* process involving the clustering of ions or molecules at the crystal surface in a similar manner to that which occurs in solution or at a foreign interface in primary heterogeneous nucleation. The key point is that the surface of the growing crystal contains active sites of higher binding energy, which drive the further incorporation of ions into the solid phase. Without these sites, a crystal would hardly grow at all under conditions of low supersaturation.

The main types of active sites are *steps* and *kinks* in the surface. The kink sites have higher binding energy than the steps because they have three 'faces' in contact with the crystal surface (Fig. 4.6). Another way of describing this is that the coordination geometry and stereochemical requirements of ions occupying kink sites are less complete—lots of 'dangling' bonds—so

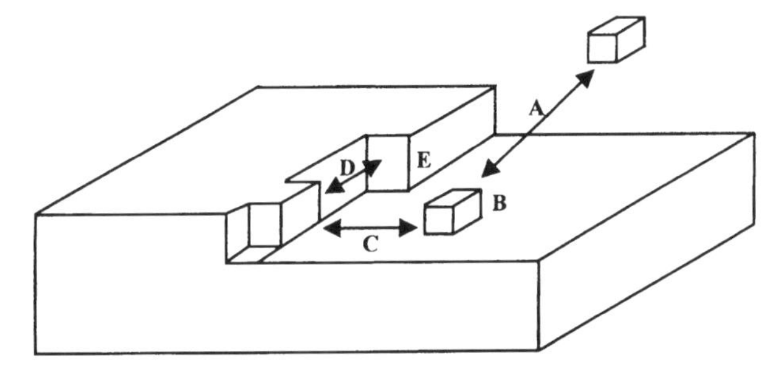

Fig. 4.6 Layer-by-layer mechanism of crystal growth. See text for details.

the surface binding of aqueous ions is higher and more energy is released than at the step sites. Crystal growth is therefore faster on those crystal faces that are most heavily covered in kinks, although it is often the case that only a small number of kinks persist at any given time.

The classical model of crystal growth assumes that ions adsorb onto the surface and diffuse to the kink sites through a series of consecutive process, such that the crystal grows by sequential addition of ions to the kink sites. As shown in Fig. 4.6, the mechanism includes:

- bulk diffusion of ions from solution to the crystal surface (stage A)
- surface adsorption and dehydration of ions on the crystal terraces (B)
- two-dimensional diffusion across the surface to the steps (C)
- one-dimensional diffusion along the step to the kink site (D)
- incorporation into the kink site (E).

Overall, the process is highly choreographic, with the ions bombarding the flat terraces of the crystal surface where they skate around losing water until pinned at the step, and are then shunted sideways into the kink site and incarcerated. Each stage involves an energy barrier and the relative magnitude of these has a profound influence on the kinetics of growth (Fig. 4.7). For example, the bulk diffusion of ions to the crystal surface is often around $6kT$ (k = Boltzmann constant, T = temperature) whereas the dehydration of ions at the crystal surface can be of the order 15 to 25 kT. Thus the removal of coordinated water molecules from metal ions adsorbed at the crystal surface is often an important aspect of the growth mechanism.

In general, the above mechanism results in the addition of ions to the kink sites so that the step advances by one growth unit along the crystal surface. Eventually the kink site will work itself out as it reaches the crystal edge, so the propagation of the step across the surface requires new kink sites to be continually produced. This doesn't cost too much energy compared with the large energy barrier that needs to be overcome when the surface layer is finally completed. At this stage, a new surface nucleation site is required to get the next layer underway and the model predicts that this depends linearly on supersaturation such that moderate levels of supersaturation, in the range of 25 to 50 per cent, are necessary. Under these conditions, the crystal grows uniformly with first-order kinetics by a *layer-by-layer* mechanism.

At higher supersaturations, multiple two-dimensional surface nuclei are formed on the crystal surfaces, which then become covered with an array of

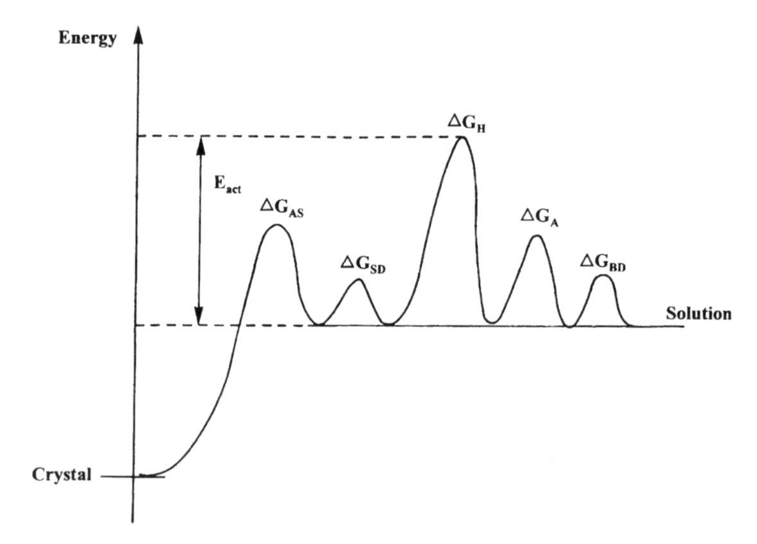

Fig. 4.7 Energy barriers (ΔG) in crystal growth. E_{act}, activation energy; AS, active site incorporation; SD, surface diffusion; H, dehydration; A, adsorption; BD, bulk diffusion.

growth islands that spread by further incorporation of ions into the kink sites. Moreover, the *polynucleation* of new islands on existing islands leads to a highly complex situation as the number of active sites increases both in and on the surface layer. This 'birth and spread' mechanism leads to a much stronger dependence on the level of the supersaturation than in the layer-by-layer process.

It turns out that many crystals grow, albeit slowly, at supersaturations as low as 1 per cent. As this is too low for the step–kink mechanism, other processes must be operating. The *screw-dislocation theory* of crystal growth accounts for these observations by considering real crystals to contain faces intersected by dislocation ledges that are sites for further crystal growth. These defects are now routinely imaged by atomic force microscopy. The ledges arise by dislocation of the lattice planes around a fixed point so they have a rotational property which means that the wedge-like step not only spreads over the surface but also winds up in a closed spiral away from the surface as further ions are added (Fig. 4.8). Because the ledge is self-perpetuating, the need for surface nucleation in the layer-by-layer mechanism is obviated and much lower supersaturation levels can be tolerated. However, the screw-dislocation mechanism is dependent on both the number of spiral steps and their rate of lateral movement over the surface, and because these are each dependent on the level of supersaturation, the rate of growth is second order for low values of the absolute supersaturation.

In reality, inorganic crystal growth from pure solution is a complex phenomenon, involving different concentrations of ions, ion-pairs and hydrated species at the crystal surface, and the observed kinetics often deviate from the idealized models described above. There may be different mechanisms operating at the same time with similar rates of growth so that the rate-limiting step will be the fastest of these competing processes. For example, the screw

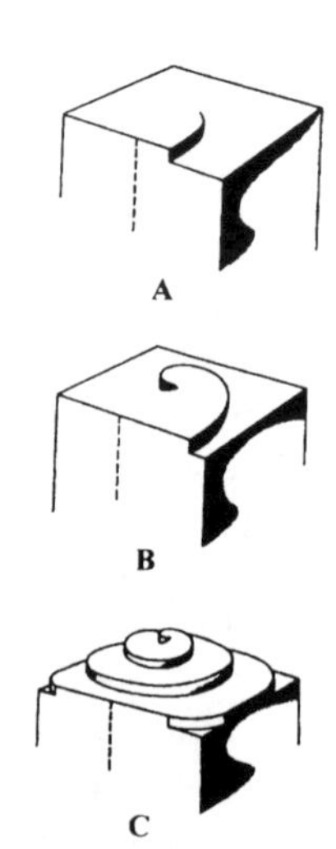

Fig. 4.8 Screw dislocation mechanism of crystal growth. Time sequence: A, B and C.

dislocation and surface nucleation mechanisms act in parallel so the fastest process will control the overall growth. Mass transport and surface adsorption processes, on the other hand, are consecutive processes with the slowest mechanism determining the rate of growth. Further complications arise because the rate-determining step may change with time as the ionic concentrations, number of defects and relative dimensions of the crystal faces become modified during crystallization.

Crystal growth from pure solution involves mass transport, surface adsorption and integration at active sites such as screw dislocations, steps and kinks. The rate of growth is dependent on the supersaturation and the number and type of active sites.

4.7 Crystal growth inhibition

Many biological fluids are supersaturated with respect to certain inorganic minerals yet crystals do not spontaneously form under normal circumstances. For example, as shown in Fig. 4.1, saliva is supersaturated with respect to hydroxyapatite precipitation but clearly our teeth do not continue to grow with time. The overgrowth is prevented by a suite of *phosphoprotein* macromolecules that are present in the fluid and bind to the surface of the enamel crystals. In general, many different types of soluble additives, such as ions, organic molecules, macromolecules and polymers, present in the crystallization solution can block the incorporation of mineral ions into the crystal surface by becoming anchored at the kink and step sites. This interference gives rise to the inhibition of crystal growth and changes in the properties and morphology (see Section 4.8) of the crystal. At very high additive levels, the propagation of steps across the crystal faces is inhibited to such an extent that the steps bunch up on themselves and the surfaces become irregular and crystal growth ultimately grinds to a halt (Fig. 4.9).

The surface layers of the growing crystal can incorporate soluble additives into the bulk lattice provided there is a high degree of complementarity in charge, size and polarization between the guest ions and vacant sites in the

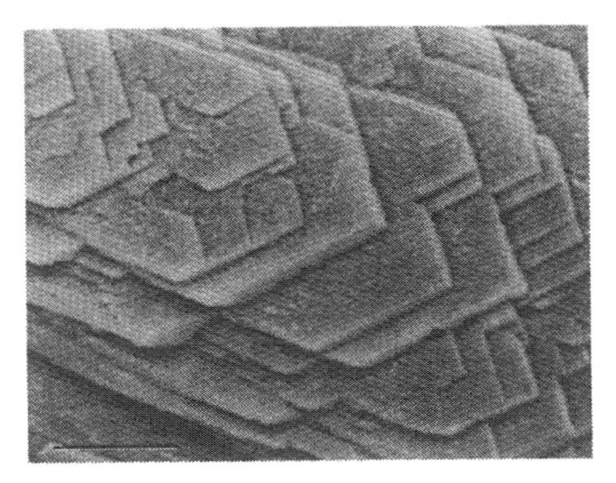

Fig. 4.9 Calcite crystal grown in the presence of high levels of aspartic acid showing bunches of steps. Scale bar, 5 μm.

boundary layers of the crystal. Isomorphic replacements, for example Mg^{2+} for Ca^{2+} in calcite and hydroxyapatite (HAP) and F^- for OH^- in HAP, usually result in small changes in the lattice spacings but significant changes in the growth rate. For example, as shown in Fig. 4.10, the reduction in pH associated with the precipitation of HAP (curve B) can be used to assess the effect of additives on the rate of crystallization. The pH change arises from the release of H^+ ions as HPO_4^{2-} ions are incorporated into the crystal structure as PO_4^{3-}. Concentrations of Mg^{2+} ions as low as 1 mM almost stop the crystallization process (curve A) due to the disruption caused by replacement of Ca^{2+} ions from lattice sites. In contrast, similar levels of F^- seem to increase the rate of crystal growth (curve C) but this is somewhat misleading because a distinct new phase, fluoroapatite, with lower solubility is precipitated in place of HAP.

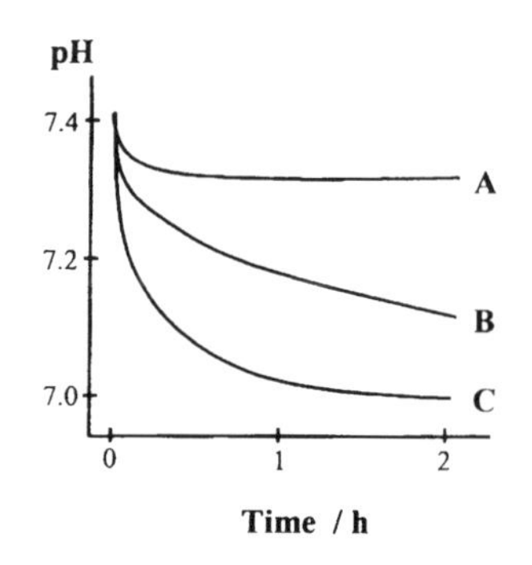

Fig. 4.10 Effects of (A) Mg^{2+} and (C) F^- ions on HAP crystallization. Curve B, control (no additives).

Because the incorporation of extraneous ions is very sensitive to the nature and distribution of the surface binding sites, chemical segregation can occur such that the guest ions becomes selectively incorporated into some crystal faces and not others. With further growth, these ions become occluded along specific crystallographic directions to give large single crystals with sectors of different composition. Similarly, extraneous ions of different charge can be accommodated but this often results in more significant and complex modifications to the host lattice. For example, there is a field of research devoted to understanding how CO_3^{2-} (and other anions) substitutes for PO_4^{3-} in the hydroxyapatite lattice. This can only occur if electroneutrality is maintained, and because each substitution results in a net gain of +1 in charge, this has to be offset either by loss of Ca^{2+} ions from lattice sites or by co-substitution with another anion such as Cl^-.

It has been known for some time that large soluble organic molecules can also dramatically inhibit the rate of growth of inorganic crystals through surface binding at the active sites. Indeed, many of the soluble macromolecules extracted from biomineralized structures show this behaviour when added to crystallization experiments performed in the laboratory. For example, calcium carbonate precipitation is associated with a corresponding decrease in pH as HCO_3^- ions are deprotonated to CO_3^{2-} prior to lattice incorporation, and the rate of crystal growth is strongly inhibited in the presence of polysaccharides isolated from the *coccoliths* of marine algae (Fig. 4.11). In general, the level of crystal growth inhibition not only depends on the amount of macromolecular additive in the surrounding supersaturated solution—often in the 0.1 mg ml^{-1} range—but also on the strength of binding to the crystal surface. The latter is dependent on the types of chemical functionality, their number and availability for cooperative binding. For instance, *amelogenins* (proteins extracted from tooth enamel, see Chapter 6, Section 6.4) with different molecular masses have different effects on the inhibition of hydroxyapatite crystallization (Table 4.4).

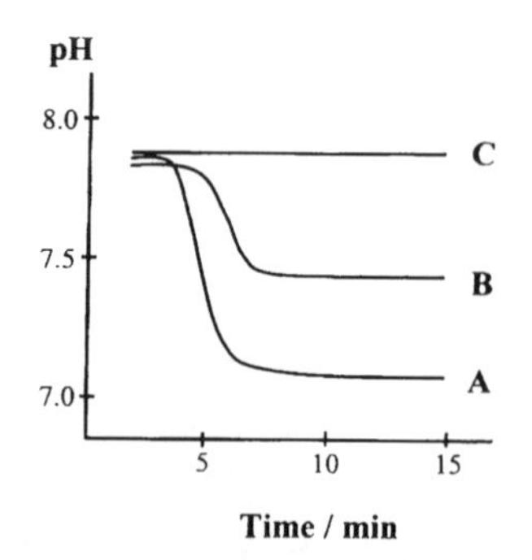

Fig. 4.11 Calcite crystallization: (A) no additives; (B) coccolith polysaccharide added after 4 min; (C) polysaccharide added at beginning of experiment.

Clearly, large organic molecules such as proteins and polysaccharides cannot be accommodated in the lattice sites of an inorganic crystal. But it is now clear that some biological macromolecules not only inhibit crystal growth but also become *intercalated* into single crystals of the inorganic mineral. For example, high-resolution X-ray diffraction investigations of the

Table 4.4 Effect of amelogenin proteins on the growth inhibition of hydroxyapatite crystals

Molecular mass	Amount used (mg)	Amount adsorbed (mg m^{-2})	% Inhibition
25 000	0.33	0.26	22
	0.66	0.47	33
20 000	0.42	0.03	<5
	0.64	0.04	5

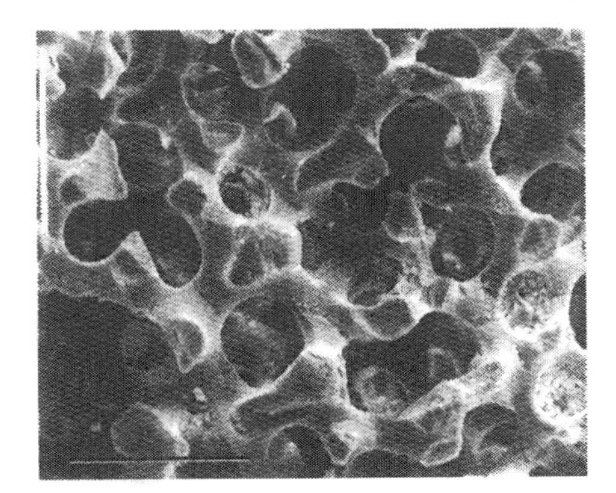

Fig. 4.12 Fractured shell of a mature sea urchin. Scale bar, 50 μm.

Mg-calcite spines of an adult *sea urchin* show that although each spine has an elaborate (non-crystallographic) surface texture and porosity, the structures are highly ordered single crystals containing 0.02 wt% of protein (see further reading). The presence of the organic intercalate has a marked effect on the mechanical properties, such that the spines fracture conchoidally like a piece of glass, and not along the low-energy cleavage planes of the Mg-doped calcite unit cell (Fig. 4.12).

But where are the macromolecules? It turns out that even a single crystal is not perfect and is made up of very slightly misoriented domains perhaps a micrometre or so in size. The boundaries between these domains are thin but quite extensive so there is sufficient space to lock in the large organic molecules during growth (Fig. 4.13). It seems possible that many crystalline biominerals are structured in this way, that is in the form of a *nanocomposite* of inorganic and organic constituents, which together produce properties not present in the individual components.

Crystal growth is inhibited in the presence of soluble additives and the crystal composition, structure and shape are often modified.

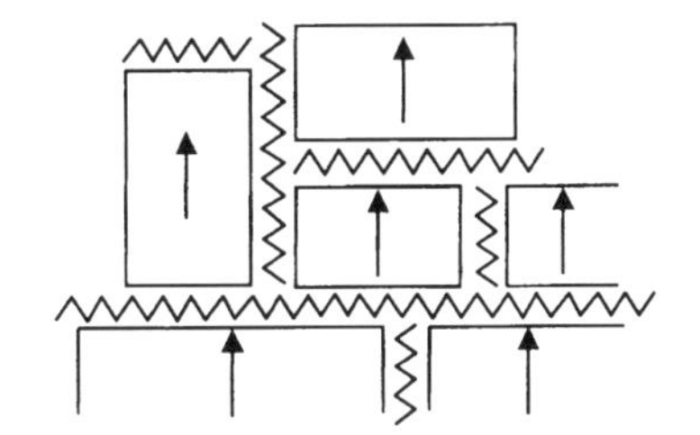

Fig. 4.13 Domain structure in a single crystal of an inorganic mineral with intercalated organic macromolecules at the coherent interfaces.

4.8 Crystal morphology

An important consequence of growth inhibition in the presence of certain additives is that the crystal morphology (*habit*) can be dramatically changed. For example, small amounts of aqueous Fe^{II} ions produce plate-like crystals of hydroxyapatite, whereas sugar molecules such as glucose give needle-

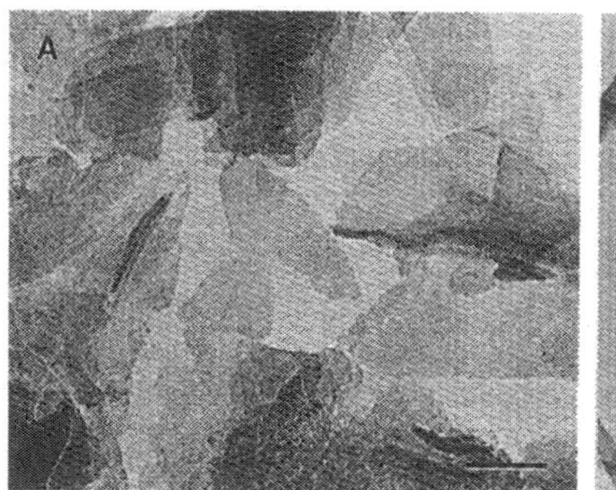

Fig. 4.14 Hydroxyapatite crystals: (A) plate-shaped; (B) needle-like morphology. Scale bars, 100 nm.

shaped particles (Fig. 4.14). In this section we discuss the factors that give rise to specific crystal shapes and how these can be modified by surface processes.

The geometric shape of a crystal is defined by the set of faces expressed to the external environment. In general, the different faces of a crystal are distinguished according to how they intersect geometrically with the axes of the unit cell. This is summarized by the *Miller index*, which is a standard method for labelling crystal planes (see further reading). Usually the index comprises three integers, *hkl*, and is written as {hkl} or (hkl) depending on whether one is referring to a general set of symmetry-related faces or a specific unique face, respectively. For example, the faces of a cube have indices (100), (010) and (001) corresponding to those lying perpendicular to the x, y and z axes of the unit cell, respectively, but can be grouped together as {100} because they are equivalent in terms of symmetry.

Because every crystal face corresponds to a unique crystallographic axis that lies perpendicular to each surface, it follows that the relative rates of growth along these directions will result in the adoption of specific geometric morphologies. For example, in Fig. 4.15 the growth of a crystal with cubic symmetry proceeds faster along the crystallographic axis referred to as [110] than it does along the [100] and [010] directions. (Note that crystallographic axes are labelled with a related index written as [*uvw*] or <*uvw*> for specific and general conditions, respectively.) This has the effect that the corresponding crystal faces that lie perpendicular to these axes change in their relative proportions as the crystal grows. In particular, the (110) face 'grows out' and

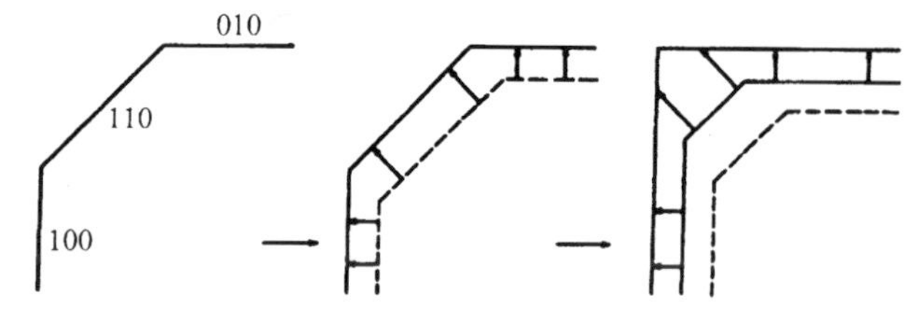

Fig. 4.15 Changes in crystal morphology due to face-specific growth rates in a cubic system.

disappears because the rate of growth is fastest in a direction perpendicular to this surface.

The key point then is that slow-growing faces are expressed in the crystal habit whereas their speedier counterparts are lost to the world. Thus, fast growth along one axis alone gives rise to a needle-shaped crystal, whereas fast growth along two directions produces a plate-like morphology and equal rates of growth in all directions yield an isotropic habit such as a cube or an octahedron. In each case, the faces perpendicular to the fast directions of growth have smaller surface areas, and the slow-growing faces dominate the morphology.

Why do the rates of growth vary on different crystal faces? We described above (Sections 4.6 and 4.7) how the rate of growth depends on surface processes that are highly sensitive to extraneous ions in the environment. Faces have different surface structures because different parts of the underlying lattice are exposed to the environment. They therefore grow at different rates because the number of kinks, steps and screw dislocations, as well as the energy associated with surface attachment and incorporation of ions, vary accordingly. Moreover, these factors are influenced to different degrees in the presence of soluble additives, so marked changes in habit can occur even for low concentrations of extraneous ions and molecules. We can only understand how this comes about by understanding the reasons that give rise to the equilibrium shape of an inorganic crystal, which we discuss in the next section.

Crystal morphology (habit) is determined by the relative rates of growth of different crystal faces, with the slow-growing surfaces dominating the final form.

4.8.1 Equilibrium morphology

Many crystals grow such that the faces comprising the habit correspond directly with the most energetically stable atomic planes in the lattice. These planes usually have low Miller indices and are related to each other according to the symmetry of the lattice (cubic, hexagonal, rhombohedral, etc.), so the crystal looks like a macroscopic version of the unit cell (Fig. 4.16). Low index faces are often stable because they contain densely packed arrays of strongly bonded atoms. However, this simple structural approach does not always apply and a more complete description, involving knowledge of the *surface energies* ($\sigma_{s\{hkl\}}$) for all the crystal planes with Miller indices {hkl}, is usually required. The surface energy of a given plane is defined as the *excess energy per unit area* of the surface lattice layer compared with the same plane in the bulk lattice. Thus, a face with low surface energy will grow slowly and contribute to the crystal habit, while the high surface energy planes become eliminated.

Fig. 4.16 Calcite crystals with rhombohedral equilibrium morphology. The unit cell has a rhombohedral space group. Scale bar, 10 μm.

If the values for $\sigma_{s\{hkl\}}$ can be obtained, either experimentally or from calculations, then the *equilibrium crystal morphology* can be predicted on the basis that the crystal habit should possess minimal total surface energy for a given volume, that is

$$\sum_{hkl} \sigma_{s\{hkl\}} \cdot A_{\{hkl\}} = \text{minimum}$$

where $A_{\{hkl\}}$ is the surface area of the {hkl} crystal face.

Because each face corresponds to a crystallographic axis that lies perpendicular to the surface, and as the rate of growth along this direction is proportional to the surface energy, then the shape of a crystal can be defined by a series of vectors of length $l_{\{hkl\}}$, each of which is proportional to $\sigma_{s\{hkl\}}$ and aligned normal to the corresponding {hkl} face. So if the surface energies of various planes are known, then it is relatively simple to use a computer program to draw out the morphology from a set of vectors of known lengths and interangular relationships (Fig. 4.17).

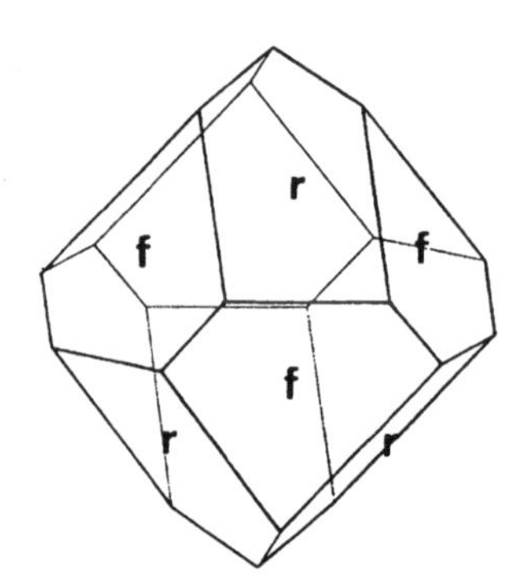

Fig. 4.17 Computer drawn morphology of a calcite crystal with rhombohedral (r) and prismatic (f) faces.

Using this approach, there have been several attempts to predict the equilibrium shape. A full treatment requires knowledge of the surface structure and a detailed description of the bonding between atoms in the surface layers and bulk lattice, so the models have until recently been semiquantitative. Initially, a lot of progress was made using the *periodic bond chain* method that dissects the surface structures into different types of 'bond chains' that exist in the planes of different crystal faces. If there are numerous chains of strongly bonded atoms in the plane of the surface then the face is stable with a low surface energy. The model uses a concept called the *attachment energy*, which is the energy per molecule released when one slice of thickness $d_{\{hkl\}}$ (d is the interplanar distance) is added onto the existing crystal face. If this value is low then the out-of-plane bonding is weak and the face grows slowly. By comparing values calculated for various different faces a ranking is drawn up from which the crystal habit can be predicted.

The periodic bond chain approach is child's play compared with recent attempts that use *atomistic simulation* to predict the equilibrium crystal morphology. For this, one needs to have a detailed understanding of the force field that describes the bonding in the crystal structure. In practice, this means that the calculation of the lattice energy has to take into account not only the electrostatic interactions between all the ions (Born–Mayer potential) but the second-order short-range repulsive and attractive forces (Buckingham potential), as well as terms for bending and torsion. The surface energies are calculated by cleaving the crystal structure along a specific direction (in virtual reality and not with a hammer) and allowing atoms on the new {hkl} surfaces to relax until they achieve a minimum energy configuration. The energy for the surface lattice is then calculated using the interatomic potentials derived for the bulk lattice along with the structural details obtained by energy minimization. The final stage involves the calculation of the surface energy for each face and is given by the excess energy obtained from the difference between the bulk and surface lattice energies.

The equilibrium crystal morphology consists of the set of symmetry-related faces that give the minimum total surface energy and can be predicted from knowledge of the surface structures and their bonding interactions.

4.8.2 Habit modification

The addition of extraneous ions and molecules to a supersaturated solution often leads to a marked change in the crystal morphology. These changes can usually be rationalized in terms of the binding of the additive to specific crystal faces, which influences the growth rates accordingly. For example, in the presence of urea, $OC(NH_2)_2$, the cubic habit of NaCl is transformed into an octahedron as the concentration of the additive is increased. Urea adsorbs specifically to the {111} planes and inhibits the growth of these faces. Initially the cubic habit is modified to a truncated cube, then a cubo-octahedron and finally an octahedron (Fig 4.18). Increasing the urea concentration even further produces a complex geometric habit in which other faces are expressed and finally irregular-shaped particles are deposited. Because it is generally true that the crystal face perpendicular to the direction of growth inhibited by the additive increases in surface area, we can read out from the changes in the relative areas of the crystal faces which crystallographic planes are directly affected by the soluble additive. This at least gives us a starting point for developing a structural model to understand how the additive interacts with the crystal surface.

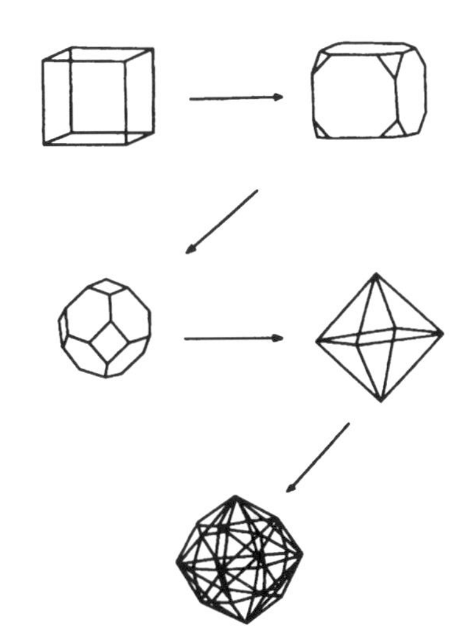

Fig. 4.18 Changes in the morphology of NaCl crystals in the presence of urea.

From our discussion in Sections 4.8 and 4.8.1, we can consider *habit modification* in terms of a change in the relative order of the surface energies due to the preferential adsorption of the additive. The exact nature of this change, however, is often difficult to unravel. For example, it might be due to selective stabilization of the surface lattice energy (thermodynamics), or mechanistic effects (kinetics) arising from changes in the numbers and type of active sites such as kinks and steps, or both. If the rate of crystal growth can be equated with the surface energies of particular crystal faces then it should be possible to accommodate the thermodynamic principles described above for the equilibrium crystal morphology within a kinetic approach to habit modification. We would then be able to calculate the relative changes in surface energies of particular faces arising from their interactions with additive molecules and predict the resulting changes in crystal morphology. Some progress has been made in recent years using this approach. For example, the surface energies for various calcite crystal faces with 50 to 100 per cent coverage of Li^+ or HPO_4^{2-} ions have been calculated and the resulting habits—hexagonal plates and elongated prisms, respectively—were predicted in agreement with those observed in experiments on calcite crystallization (Fig. 4.19). For more details, see further reading.

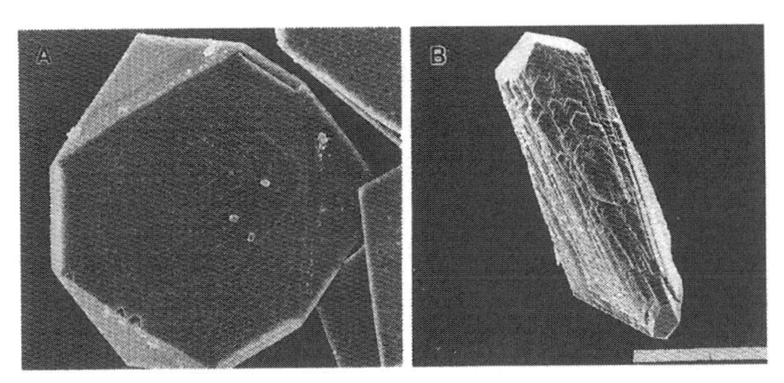

Fig. 4.19 Calcite crystals grown in the presence of: (A) Li^+, plate with hexagonal {001} faces and rhombohedral {104} side faces; (B) HPO_4^{2-}, prism with rhombohedral {104} top faces and $\{1\bar{1}0\}$ side faces. Scale bars, 10 μm and 20 μm, respectively.

There is a mounting body of circumstantial evidence that strongly suggests that the principal cause for the face selectivity of certain additives is *molecular recognition*. This implies that the surface interactions are highly specific because of a high degree of charge, stereochemical and structural matching between the additive and growth sites on the crystal surface. For example, α,ω-dicarboxylic acids with the general formula $(CH_2)_n(CO_2H)_2$ (Table 4.5) are highly effective at stabilizing the prismatic $\{1\bar{1}0\}$ faces of calcite provided that both carboxylate groups are ionized and $n < 3$. These faces lie parallel to the crystallographic c axis and contain both Ca^{2+} and CO_3^{2-} ions, with the plane of the triangular anion perpendicular to the surface such that two oxygen atoms are exposed (Fig. 4.20). The short-chain additives interact specifically with this surface because the stereochemical arrangement is maintained on dicarboxylate binding, and both carboxylate groups can bind simultaneously to two different calcium ions if the spacing is close to 0.4 nm. Malonate and maleate fit this criterion but the increased rigidity of the latter reduces the binding affinity. In contrast, fumarate has no morphological effect because the *trans* stereochemistry prevents both carboxylate groups from interacting with the surface in a cooperative fashion. The potency and morphological specificity of these dicarboxylate additives are lost at high concentrations where non-specific binding becomes paramount. On the other hand, they can be increased by additional charge functionalization in the molecule. For example, both α-aminosuccinate (aspartate) and γ-carboxyglutamate (see Table 4.5) show more effective stabilization of the prismatic calcite faces than succinate or glutamate, respectively (Fig. 4.21).

Fig. 4.21 Prismatic calcite crystal formed in the presence of γ-carboxyglutamate. Scale bar, 10 μm.

Other experiments indicate that *macromolecular additives*—including proteins extracted from biomineralized structures—can have highly specific effects on the morphology of inorganic crystals. For example, polysaccharides such as sodium alginate and various carragheenans extracted from

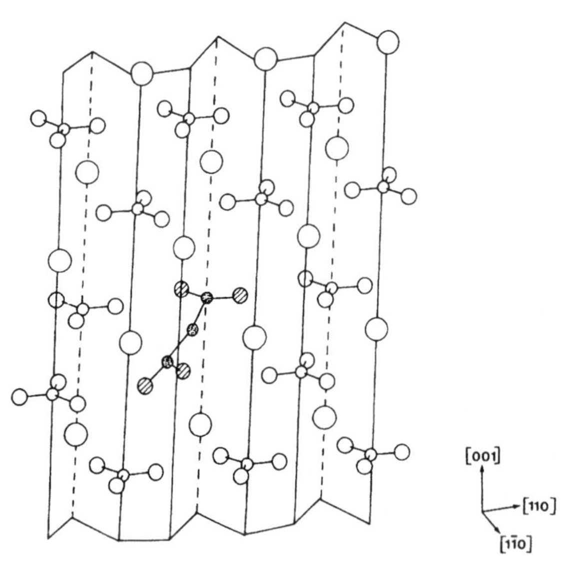

Fig. 4.20 Drawing of calcite $\{1\bar{1}0\}$ crystal face with surface-adsorbed malonate anion.

Table 4.5 Organic dicarboxylates used as habit modifiers in calcite crystallization. From top to bottom: column A, oxalate, malonate, succinate and glutarate; column B, maleate and fumarate; column C, aspartate and γ-carboxyglutamate

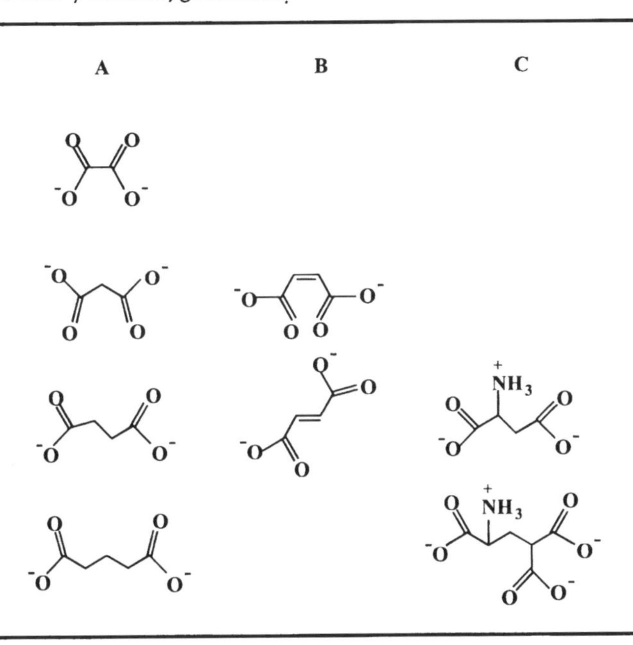

seaweed (Fig. 4.22), when added to supersaturated solutions of sodium chloride, inhibit surface nucleation by adsorption onto edge sites allowing dislocation growth to dominate over edge nucleation. The resulting crystals have a well-defined cubic habit compared with the control crystals prepared in the absence of the inhibitors. Clearly, when one begins to consider the stereochemical possibilities of macromolecular interactions with inorganic crystal faces, then the recognition processes become extremely complex. Studies have shown that acidic macromolecules extracted from the Mg-calcite biominerals of adult *sea urchins* interact specifically with the prismatic faces of calcite crystals grown in the laboratory (see further reading). These molecules have high levels of aspartic and glutamic acid residues, suggesting that the acidic amino acid carboxylate groups behave in a similar stereochemical fashion as observed with the low molecular weight dicarboxylate additives described above. Indeed, computer models have shown that synthetic analogues of these proteins, such as polyaspartate with 11 repeating carboxylate units, $[Asp]_{11}$, interact strongly with the prismatic faces of calcite (Fig. 4.23). Details on the use of polyaspartate in inorganic crystallization can be found under further reading.

Low and high molecular weight additives—including proteins extracted from biominerals—can induce habit modifications by changing the relative growth rates of different crystal faces through molecular-specific interactions with particular surfaces. Electrostatic, stereochemical and structural matching are important factors that significantly modify the surface energy or mechanism of growth, or both.

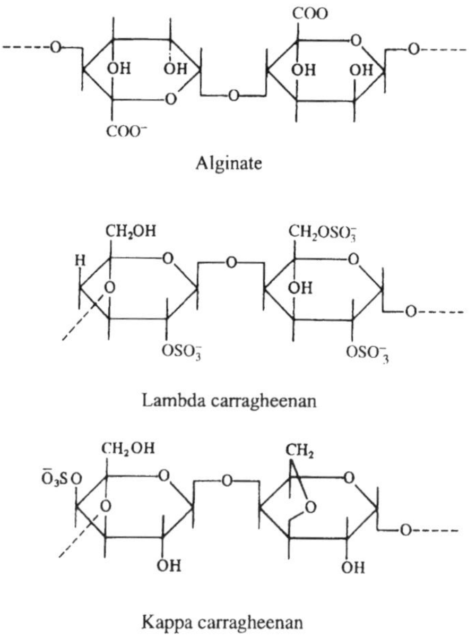

Alginate

Lambda carragheenan

Kappa carragheenan

Fig. 4.22 Various polysaccharides.

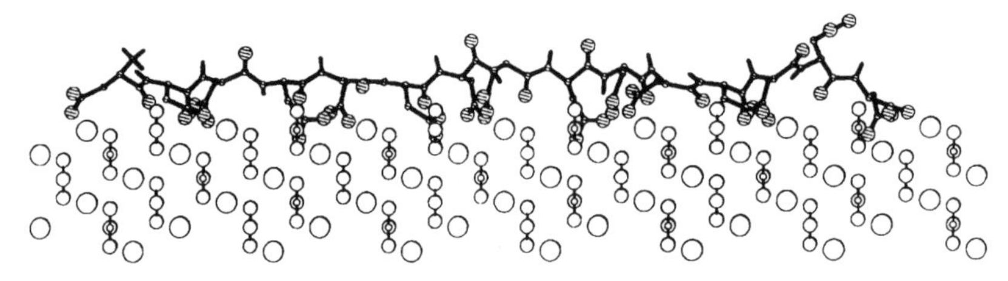

Fig. 4.23 Computer model showing side view of the calcite {1$\bar{1}$0} face with surface-bound polyaspartate ($[Asp]_{11}$).

4.9 Polymorphism

Amazingly, proteins extracted from either the calcite-containing prismatic layers or the aragonite nacreous layers of seashells induce the crystallization of calcite or aragonite, respectively, when added to supersaturated solutions of calcium bicarbonate in the laboratory (see further reading). Both minerals are made of $CaCO_3$ but clearly the selection of different unit-cell structures indicates fundamental changes in the crystallization process.

From the perspective of chemical control, the selection of a particular polymorph can arise by kinetic effects that influence the nucleation and growth pathways. (Another possibility, which involves the use of interfacial templates to control the structure, is discussed in Chapter 6, Sections 6.7.5 and 6.7.6.) For example, it is well known that at high concentrations, Mg^{2+} ions will switch the crystallization of calcite to aragonite precipitation. At first sight this seems counterintuitive because Mg^{2+} is known to isomorphically replace Ca^{2+} ions in calcite but cannot enter the aragonite lattice. However, this thermodynamic condition is offset by the kinetic effects arising from the interaction of Mg^{2+} ions with small crystals and nuclei of calcite, which disrupts the surface and reduces the rate of crystal growth. At the same time, aragonite nuclei, which are not affected by the additive, continue to grow unabated in the supersaturated solution and therefore become the dominant polymorph in the crystallization process.

Polymorph selectivity can also arise from a transformation process that starts with an amorphous mineral and proceeds through a series of structures with the same composition but with increasing thermodynamic stability. The basic principle that predicts the sequence of polymorphs produced in crystal growth is based on an empirical observation called the *Ostwald–Lussac law of stages*. This principle states that under conditions of sequential precipitation, the initial phase formed is the one with the highest solubility followed by hydrated polymorphs and then a succession of crystalline phases in order of decreasing solubility. We should therefore be able to read off from a table of solubility products, such as those listed in Table 4.1, and predict the order of deposition for a crystallization system involving several polymorphs. This is clearly a kinetic argument because according to our previous discussions, the supersaturation level for the least soluble polymorph will be larger than the less stable crystalline or amorphous phases, so thermodynamically we should expect the most stable polymorph to be the first to precipitate. The preferential formation of the most soluble polymorph therefore implies a faster rate of nucleation (J_N) for the less stable structures. As shown in Fig. 4.24, although the supersaturation for an amorphous phase (S_{amorph}) is less than that for a thermodynamically stable crystalline phase ($S_{crystal}$) at a given value of the activity product, the rate of nucleation of the disordered structure is higher. From the nucleation equations discussed in Section 4.4, we can see that this can occur if the lower supersaturation level is offset by a significantly reduced value for the interfacial energy (σ). This is usually the case for the nucleation of an amorphous phase or polymorphs with less stable crystalline structures.

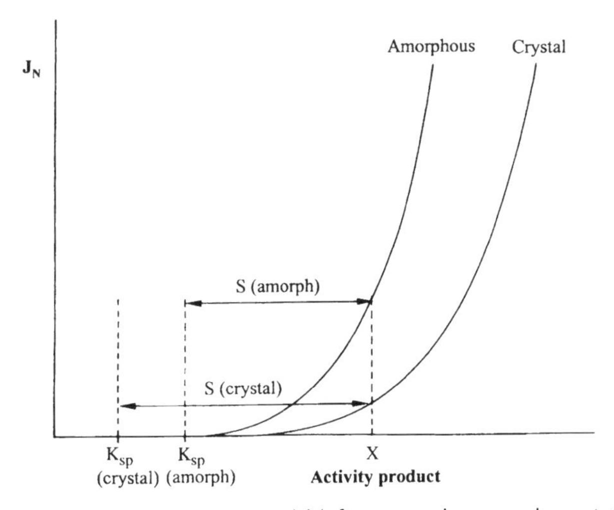

Fig. 4.24 Plot of nucleation rates (J_N) for amorphous and crystalline polymorphs as a function of activity product. Values of supersaturation (S) and solubility product (K_{sp}) are shown.

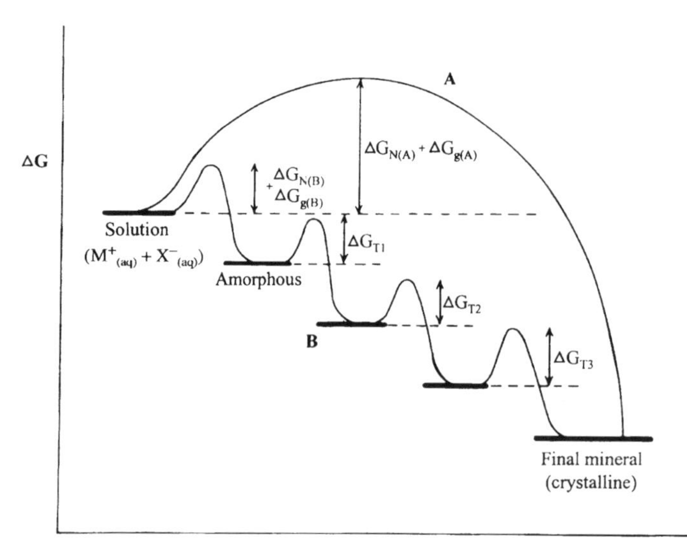

Fig. 4.25 Pathways to crystallization and polymorph selectivity: (A) direct; (B) sequential. See text for details.

As shown in Fig. 4.25, whether a system follows a one-step route to the final mineral phase (pathway A) or proceeds via sequential precipitation (pathway B), depends on the activation energies associated with nucleation (N), growth (g) and transformation (T). One of the most important factors revolves around the actual structure of the critical nucleus because this determines how the embryonic state is modified during the course of crystal growth. Unfortunately, there is little structural information about the initial states formed in precipitation processes because the clusters are transient and of nanoscale dimension.

In principle, we can envisage a large number of nuclei with different structures that depend on the strength and nature of the interactions between ions sited on regular lattice positions, and ions and solvated species randomly located within the critical nucleus. Although there are countless variations, one end-member of this series of structures would resemble a small piece of the bulk crystalline phase (Fig. 4.26A). This model of the nucleus involves strong interactions between the ions so that the lattice energy is high enough to overcome solvation and offset the surface energy. Although the ions in the critical nuclei are relaxed to some degree from their normal unit-cell positions, there is still a close correspondence between the lattice parameters of the nucleus structure and that of the bulk phase. In this situation, nucleation and growth are structurally linked—in a sense, growth just reiterates nucleation—so that in principle, chemical control over nucleation would determine polymorph selectivity by a one-step mechanism, shown as pathway A in Fig. 4.25. Alternatively, the stable critical cluster may be best described in terms of weakly interacting hydrated ions that form a diffuse solvated solid phase with no regular structure (Fig. 4.26B). In this model, the lattice energy is low but so is the surface energy because of solvation. The amorphous phase is precipitated first, followed by a polymorphic series, consistent with the Ostwald–Lussac law and pathway B in Fig. 4.25. The amorphous phase will be favoured particularly in systems where the crystalline polymorphs have high interfacial energies and large unit-cell volumes. For instance, if the criti-

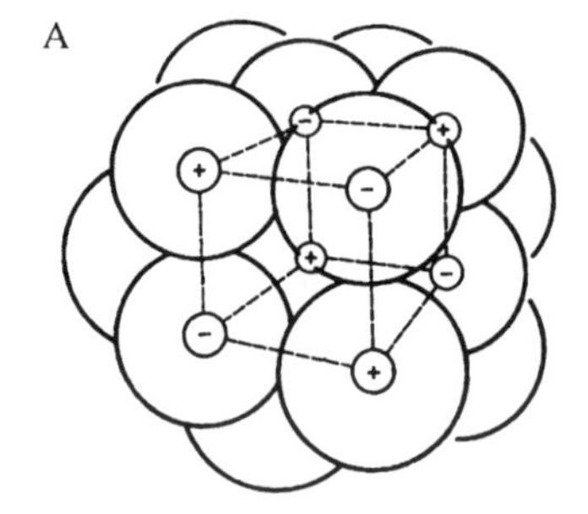

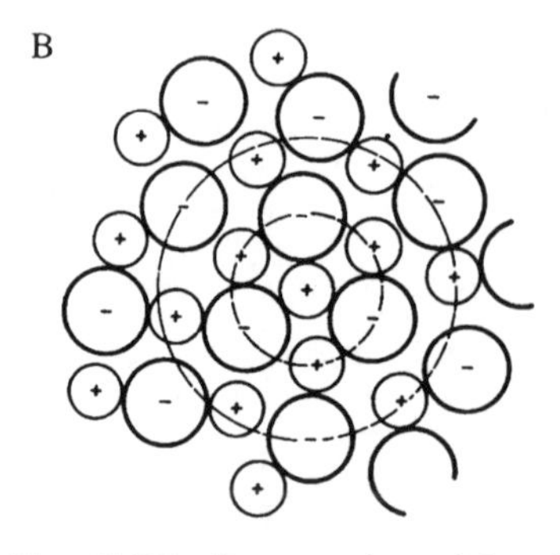

Fig. 4.26 Structural models of nuclei: (A) periodically ordered, and (B) disordered and solvated.

cal nucleus contains fewer ions than are actually required to complete the unit cell, then nucleation occurs prior to the construction of the periodic structure.

Polymorph selectivity can be chemically controlled by kinetic effects involving additives or through transformation processes that proceed along a series of structures with decreasing solubility and increasing thermodynamic stability. The structure of the critical nucleus is an important factor in controlling the crystallization pathway.

4.10 Phase transformations

Crystal growth in solution can also occur via a sequential process that involves both structural and compositional modifications of precursor and intermediate phases. The changes in composition and structure are mediated by *phase transformations* that usually occur by surface dissolution of the precursor followed by nucleation of the second phase. This often occurs directly on the surface of the initially formed particles. How far these phase transformations proceed along a pathway of intermediates depends on the solubilities of the amorphous precursor and crystalline intermediates and the free energies of activation of their interconversions. The situation is therefore very similar to that described for pathway B in Fig. 4.25 for the case of polymorphism, where the first phase precipitated is amorphous followed by a sequence of minerals with decreasing solubility. Again, the biological trick is to be able to stop the process at a particular structure that would otherwise transform into a more stable crystalline phase (or polymorph).

For example, there are many known inhibitors of hydroxyapatite formation (Table 4.6) which act by stopping the transformation of amorphous calcium phosphate to crystalline intermediates or in a few cases octacalcium phosphate to hydroxyapatite. Their inhibitory action can be overcome in the case of phosphometabolites and proteins by the addition of enzymes for which these molecules act as natural substrates. This suggests that 'promotion' of various mineral phases and polymorphs does not occur in the literal sense, but

Table 4.6 Biological inhibitors of hydroxyapatite crystallization from aqueous solution

Mg^{2+}
CO_3^{2-}
Pyrophosphate ($P_2O_7^{4-}$)
Polyphosphates
Nucleotide polyphosphates
adenosine triphosphate
guanosine diphosphate
glucose 1,6-diphosphate
Cartilage proteoglycans
Dentine phosphoproteins
Polycarboxylates
Phospholipids
Phosphocitrate

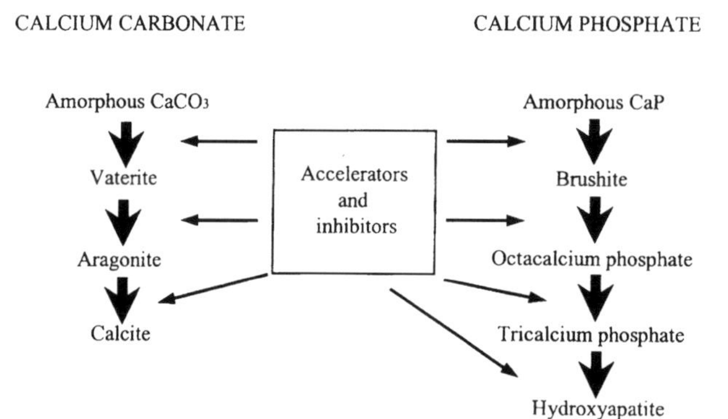

Fig. 4.27 Use of additives in the regulation of polymorph selectivity (calcium carbonate) and phase transformations (calcium phosphate).

that the mineralization pathway is controlled through intermittent release of a system under chemical repression. The crystallographic selection then becomes dependent on the extent to which the sequence of chemical inhibition and release is allowed to proceed within the mineralization site (Fig. 4.27). Moreover, modulation of the sequence in time and biological location influences the development of alternative crystallographic structures and compositions during different stages of the mineralization process.

One important consequence of this multiphase pathway is that it contrasts with other theories of biomineralization that propose a central role for organic polymeric matrices in controlling the structure of biominerals by primary nucleation (see Chapter 6, Section 6.7, for details). In reality, both processes probably operate at different times and places within the biological system.

The chemical control of crystallization pathways that involve a sequence of kinetic inhibition and phase transformation can result in a high degree of selectivity in crystal structure and composition.

4.10.1 Amorphous precursors

Amorphous phases are more soluble than their crystalline counterparts at equilibrium, so we might expect to find disordered precursors to be precipitated throughout biomineralization (see further reading). Indeed, there is growing evidence that *amorphous granules* containing high levels of inorganic and organic components are prevalent in the early stages of many biomineralization systems. These structures are often formed away from the mineralization site and subsequently transported to the organic matrix where they aggregate in large numbers before undergoing phase transformation to more stable minerals.

The problem of course is that these phases do not stay around for long and even if present in the mature structures are readily dissolved during sample preparation procedures. Only amorphous silica is stabilized indefinitely at room temperature and pressure—hence the extensive field of biological silicification—because the activation energy for transformation to quartz is very high, *ca.* 800 kJ mol^{-1}. Amorphous calcium phosphate, in contrast, transforms slowly at neutral pH to octacalcium phosphate and hydroxyapatite in aqueous media. With time, the amorphous calcium phosphate particles are consumed as they become enveloped by a crystalline mass that grows outwards from the interior of the dissolving precursor. This process is biologically relevant in the formation of certain types of bone, and in the teeth of *chitons* (molluscs) where the conversion of amorphous calcium phosphate directly to hydroxyapatite can be followed in detail by infrared spectroscopy (see further reading). Because the amorphous phase is a random network of bonds rather than a periodic structure, there is no direct structural correspondence with the crystalline phases so the phase transformations must proceed by dissolution. The process is therefore diffusion or surface-reaction controlled and highly dependent on temperature and pH.

Amorphous calcium carbonate is highly soluble and rapidly transforms unless kinetically stabilized. In biomineralization, this is achieved by enclosing the amorphous calcium carbonate in an impermeable organic sheath. In plant *cystoliths*, this is achieved with a polysaccharide-rich overlay, whereas proteins are employed in the early developmental stages of larval *sea urchin* micro-skeletons. In the latter case—see further reading for details—high levels of Mg^{2+} are accommodated in the amorphous phase so that the resulting (slow) phase transformation produces a Mg-calcite tri-radiate spicule with levels of up to 30 mol% Mg^{2+} in the crystal structure (Fig. 4.28). Such high levels are not achievable for calcite crystallization in the laboratory under ambient conditions because the concomitant amounts of Mg^{2+} ions needed in the supersaturated solution result in the kinetic precipitation of aragonite. Thus, the amorphous to crystalline pathway opens up a route to significant compositional modifications in the structure of biominerals that are not possible through direct precipitation.

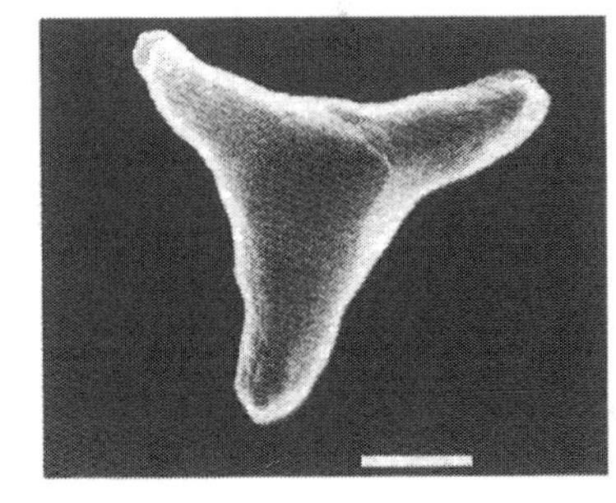

Fig. 4.28 Early growth stage in larval sea urchin micro-skeleton. Scale bar, 1 μm.

Amorphous phases are key precursors in phase transformation processes and, except for silica, transform into crystalline minerals sometimes with unusual compositions, unless protected by an organic sheath.

4.10.2 Crystalline intermediates—calcium phosphates

Phase transformation processes are of significant importance in calcium phosphate biomineralization because there are several crystalline phases with significantly different solubilities that are pH-dependent. This has an interesting effect in that the order of solubilities changes as the pH is varied. For example, as shown in Fig. 4.29, the hydrated hydrogen phosphate mineral brushite ($CaHPO_4 \cdot 2H_2O$, dicalcium phosphate dihydrate, DCPD) is the least soluble phase at pH values below 4 because phases such as hydroxyapatite (HAP) and octacalcium phosphate (OCP) readily dissolve by protonation of the PO_4^{3-} ions in the crystal lattice. Above this pH, however, the solubilities of both HAP and OCP dramatically fall so that DCPD becomes the most soluble phase at pH values above 5. Using the Ostwald–Lussac law we would therefore predict that the order of precipitation of crystalline calcium phosphates at pH values close to neutral would follow the sequence DCPD to OCP to HAP, and indeed this is often observed experimentally.

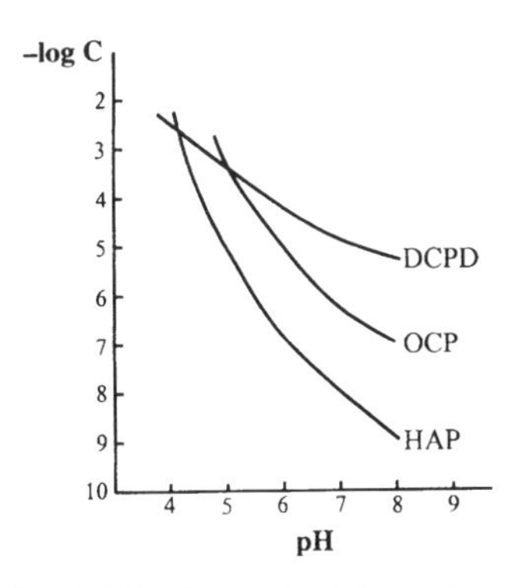

Fig. 4.29 Plot of calcium phosphate solubilities (–log *C*, where *C* = [total Ca^{2+}] x [total phosphate]) against pH for various phases. DCPD, dicalcium phosphate dihydrate; OCP, octacalcium phosphate; HAP, hydroxyapatite.

Although dissolution and surface-controlled processes often dominate the interconversion of calcium phosphate phases, *in situ* solid-state transformation does occur when there is a close structural match and low interfacial energy between the two structures. For example, OCP transforms to HAP via an *in situ* solid-state reaction involving the hydrolysis of one unit-cell thickness of OCP (interplanar spacing, $d_{(100)}$ = 1.868 nm) to give two unit-cell slices of HAP ($2d_{100}$ = 1.632 nm) and a unit-cell contraction of 0.236 nm along the *a* axis. A relict of this process is often observed in biological hydroxyapatites, such as dental *enamel*, where the HAP crystallites contain a one unit-cell-thick residual layer of OCP—referred to as the *central dark line*—which can be clearly distinguished in high-resolution electron microscopy images (Fig. 4.30). For more details on calcium phosphate crystallization, see further reading.

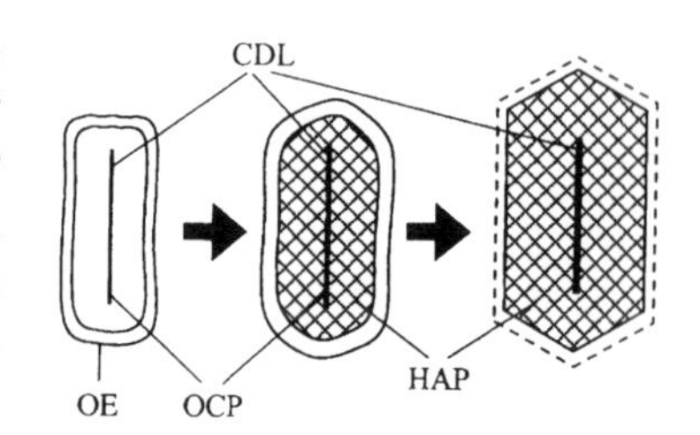

Fig. 4.30 Formation of enamel crystals. An octacalcium phosphate (OCP) precursor phase is formed within an organic envelope (OE) and then overgrown with a single crystal of hydroxyapatite (HAP). Traces of the OCP phase are left as a central dark line (CDL) inside the mature crystal.

Phase transformations are important in the biomineralization of crystalline calcium phosphates. They often occur through sequential precipitation involving surface or solid-state processes according to the Ostwald–Lussac law of stages.

4.10.3 Rusty transformations—iron oxides

Phase transformation processes are also important in the deposition of iron oxide minerals. The main phases include hematite (α-Fe_2O_3), goethite (α-FeOOH), lepidocrocite (β-FeOOH), magnetite (Fe_3O_4) and an amorphous or poorly ordered phase, called ferrihydrite ($Fe_2O_3 \cdot nH_2O$), which is the stuff of rust. All the minerals are ferric oxides (Fe^{III}) except for magnetite which is a mixed valance compound ($Fe^{III}_2Fe^{II}O_4$). The possible interconversions and their associated chemical conditions are summarized in Fig. 4.31. Ferrihydrite is of key significance because it can be transformed into each of the other phases by solid-state transformation to hematite, dissolution and slow reprecipitation to goethite, or reductive dissolution to the mixed valence oxide, magnetite. Except for hematite, which is unknown in biomineralization, all these iron oxides play an important role in biomineralization (see Chapter 2, Section 2.6). In many cases, for example in the formation of goethite and magnetite in the teeth of limpets and chitons, respectively, the ferrihydrite cores of the iron storage protein, ferritin, play an important role in the transformation processes (Fig. 4.31).

The remarkable transformation of ugly old rust to beautifully refined crystals of magnetite in *magnetotactic bacteria* has been studied in some detail

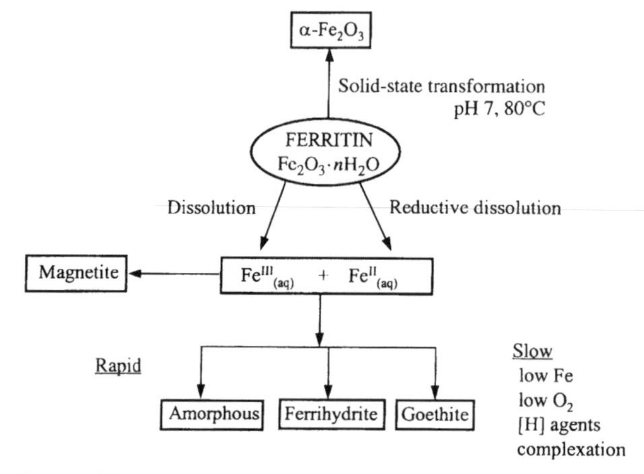

Fig. 4.31 Iron oxides and their interconversions.

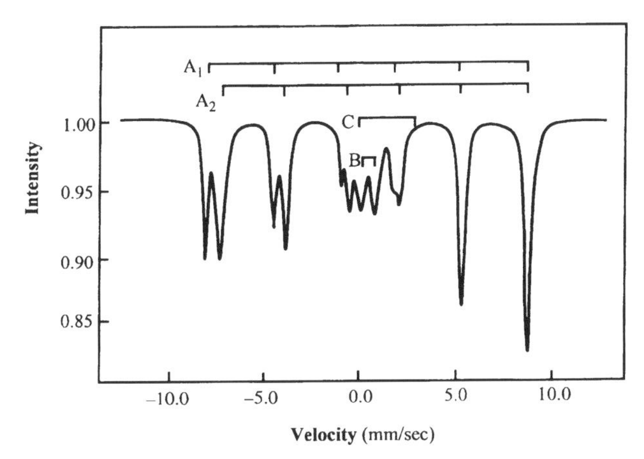

Fig. 4.32 ^{57}Fe Mössbauer spectrum of frozen magnetotactic bacteria. See text for details.

(see further reading). ^{57}Fe Mössbauer spectra clearly show peaks for well-ordered magnetite and an amorphous or poorly ordered ferrihydrite phase associated with the bacterial cells at various stages of development (Fig. 4.32). (We described the use of this technique for studying the ferrihydrite cores of ferritin in Chapter 2, Section 2.6.2.) The magnetite lattice shows up as two sets of partial overlapping sextets that correspond to Fe^{III} ions in tetrahedral sites (A_1) sites or Fe^{II} and Fe^{III} ions in octahedral sites (A_2 sites). In addition, two doublets are observed (peaks B and C in Fig. 4.32) from ferrihydrite and aqueous Fe^{II} ions, respectively. The ferrihydrite phase is closely associated with the early stages of magnetite crystallization as shown by electron microscopy images of immature crystals. At this stage, the individual particles are irregular and consist of contiguous domains of magnetite and the poorly ordered precursor but with time transform into highly ordered pure magnetite crystals.

The phase transformation of rust to magnetite is slow because dehydration, dissolution and nucleation are involved. Dissolution of ferrihydrite occurs by partial reduction of some of the surface Fe^{III} ions to Fe^{II} ions that are much more soluble. Investigations of this process in the laboratory indicate that the critical step in the phase transformation involves the binding and interaction of the aqueous Fe^{II} ions at the surface of the ferric oxide, which release protons into solution due to hydrolysis (Fig. 4.33A). Provided that the number of active sites on the surface remains approximately constant during the reaction, the overall reaction rate is first order with respect to the concentration of Fe^{II} ions in solution. The next stage involves the destabilization of the surface intermediate, resulting in dissolution and release of mixed valence hydroxo-complexes that subsequently reprecipitate as magnetite with the loss of additional protons (Fig. 4.33B). The whole process occurs at the interface between the poorly ordered and crystalline phases such that with time the precursor disappears as the magnetite crystals grow. Further details on the biological control of this system are found in Chapter 5, Section 5.1.1.

Phase transformations are important in the biomineralization of iron oxides. Ferrihydrite, sometimes in the form of the protein ferritin, is transformed to goethite in limpet teeth or magnetite in chiton teeth and magnetotactic

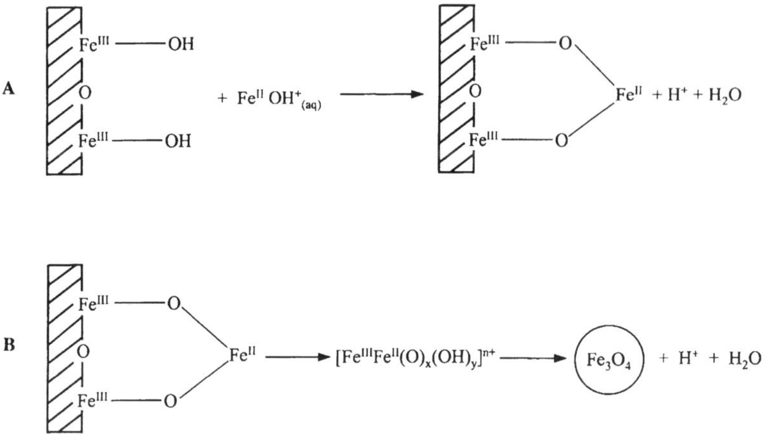

A $\equiv Fe^{III}\text{–}OH\ (\times 2) + Fe^{II}OH^{+}_{(aq)} \longrightarrow \equiv(Fe^{III}\text{–}O)_2Fe^{II} + H^+ + H_2O$

B $\equiv(Fe^{III}\text{–}O)_2Fe^{II} \longrightarrow [Fe^{III}Fe^{II}(O)_x(OH)_y]^{n+} \longrightarrow Fe_3O_4 + H^+ + H_2O$

Fig. 4.33 Transformation of ferrihydrite to magnetite: (A) surface binding of Fe^{II} aquo-species; (B) dissolution of mixed valence complexes and magnetite nucleation.

bacteria. ^{57}Fe Mössbauer spectroscopy and electron microscopy are important physical methods used to study these processes.

4.11 Summary

In this chapter we have looked at four concepts that lie behind the chemical control of biomineralization—solubility (and solubility product), supersaturation, nucleation and crystal growth. Although the solubility product determines the thermodynamic limit for the onset of precipitation, it is often difficult to apply or determine quantitatively in biological environments. How far the system is out of equilibrium depends on the supersaturation, which also governs the rate of nucleation and growth. Nucleation is also strongly influenced by the interfacial energy of the critical nucleus, which can be significantly reduced in the presence of an insoluble substrate. A high degree of lattice matching between the substrate and mineral nuclei leads to oriented overgrowth.

Crystal growth involves mass transport, surface adsorption and integration at active sites such as screw dislocations, steps and kinks. The rate of growth is often severely inhibited in the presence of additives, which also produce changes in the composition, structure and shape of the crystals. Although the equilibrium morphology of a crystal can be predicted from the surface structures and their bonding interactions, many low and high molecular weight additives modify the habit by changing the relative growth rates of different crystal faces through molecular-specific interactions involving electrostatic, stereochemical and structural matching.

The chemical control of crystallization pathways in biomineralization can result in a high degree of selectivity in the structure (polymorphism) and composition of crystalline minerals, such as calcium phosphates and iron oxides. In general, this occurs by kinetic inhibition involving additives or through phase transformation processes that proceed along a series of structures with decreasing solubility and increasing thermodynamic stability. Amorphous phases are important precursors in this mechanism and, except for silica, readily transform to crystalline minerals unless protected by organic sheaths.

With the chemical concepts in hand, we can now proceed to discuss some specific biological mechanisms that account for the controlled deposition of biominerals. We begin with boundary-organized biomineralization, which brings together the chemical regulation of supersaturation and spatial control of biological environments. The details of this process are discussed in Chapter 5.

Further reading

Belcher, A. M., Wu, X. H., Christensen, R. J., Hansma, P. K., Stucky, G. D and Morse, D. E. (1996). Control of crystal phase switching and orientation by soluble mollusc-shell proteins. *Nature*, **381**, 56–58.

Berman, A., Addadi, L. and Weiner, S. (1988). Interactions of sea-urchin skeleton macromolecules with growing calcite crystals—a study of intracrystalline proteins. *Nature*, **331**, 546–548.

Berman, A., Hanson, J., Leiserowitz, L., Koetzle, T. F., Weiner, S. and Addadi, L. (1993). Biological control of crystal texture: a widespread strategy for adapting crystal properties to function. *Science*, **259**, 776–779.

Brown, B. E. (1982). The form and function of metal-containing granules in invertebrate tissues. *Biol. Rev.*, **57**, 621–667.

Burton, W. K., Cabrera, N. and Frank, F. C. (1951). The growth of crystals and the equilibrium structure of their surfaces. *Philos. Trans. R. Soc. London A*, **243**, 299–358.

Frankel, R. B., Papaefthymiou, G. C., Blakemore, R. P. and O'Brien, W. D. (1983). Fe_3O_4 precipitation in magnetotactic bacteria. *Biochim. Biophys. Acta*, **763**, 147–159.

Garside, J. (1982). Nucleation. In *Biological mineralization and demineralization* (ed. Nancollas, G. H.), pp. 23–25. Springer-Verlag, Berlin.

Hammond, C. (1992). *Introduction to crystallography.* Microscopy Handbook (19), Royal Microscopical Society, Oxford University Press, New York.

Johnson, D. A. (1982). *Some thermodynamic aspects of inorganic chemistry.* Cambridge University Press, Cambridge.

Lowenstam, H. A. and Weiner, S. (1985). Transformation of amorphous calcium phosphate to crystalline dahlite in the radular teeth of chitons. *Science*, **227**, 51–53.

Mullin, J. W. (1971). *Crystallization*. Butterworths Press, London.

Nancollas, G. H. (1989). In vitro studies of calcium phosphate crystallization. In *Biomineralization: chemical and biochemical perspectives* (ed. Mann, S., Webb, J. and Williams, R. J. P.), pp. 157–187. VCH Verlagsgesellschaft, Weinheim.

Nielson, A. E. (1964). *Kinetics of precipitation.* Pergamon Press, Oxford.

Raz, S., Weiner, S. and Addadi, L. (2000). Formation of high-magnesium calcites via an amorphous precursor phase: possible biological implications. *Adv. Mater.*, **12**, 38–42.

Sikes, C. S. and Wierzbicki, A. (1996). Polyamino acids as antiscalants, dispersants, antifreezes and adsorbent gelling materials. In *Biomimetic materials chemistry* (ed. Mann, S.), pp. 249–278. VCH, New York.

Simkiss, K. (1991). Amorphous minerals and theories of biomineralization. In *Mechanisms and phylogeny of mineralization in biological systems* (ed. Suga, S. and Nakahara, H.), pp. 375–382. Springer-Verlag, Tokyo.

Titiloye, J. O., Parker, S. C. and Mann, S. (1993). Atomistic simulation of calcite surfaces and the influence of growth additives on their morphology. *J. Cryst. Growth*, **131**, 533–545.

Walton, A. G. (1967). *The formation and properties of precipitates*, Vol. 23. Interscience, New York.

5 Boundary-organized biomineralization

One key aspect of the biological control of mineralization is the partitioning of space either within or outside the cell by organic structures. We refer to this process as *boundary-organized biomineralization*. Because the enclosed environments are spatially defined and chemically separated from the general biochemical activity of the cells, they have several important functions, including:

- *spatial delineation*—for controlling the size, shape and organization of the mineral phase
- *diffusion limited ion flow*—for controlling ionic activities, solution compositions and supersaturation
- *mineral passivation*—for the surface stabilization of minerals against dissolution or phase transformation
- *ion accumulation and transport*—for supplying chemicals to remote intra- and extracellular mineralization sites
- *mineral nucleation*—for regulating interfacial energies (see Chapter 6, Section 6.7)
- *mineral transportation*—for moving mineralized structures to new construction sites (see Chapter 8, Section 8.2).

In this chapter, we discuss the main types of organic compartments associated with boundary-organized biomineralization and their central role in controlling the chemical composition and supersaturation levels of mineralization sites. A more detailed discussion of how these structures are used to produce minerals with complex shapes can be found in Chapter 7.

The delineation of biological environments is of key importance in boundary-organized biomineralization because it provides sites of controlled chemistry that are spatially defined.

5.1 Spatial boundaries

There are four main types of organic structures that are used in boundary-organized biomineralization:

- membrane-bounded phospholipid vesicles
- polypeptide vesicles
- cellular assemblies
- macromolecular frameworks.

In general, these enclosed environments are constructed from cooperative processes that involve reinforcing non-covalent interactions between organic molecules such as phospholipids, polypeptides, proteins and polysaccharides.

Because this invariably occurs prior to mineralization, the term *supramolecular preorganization* is used to describe the general chemical features associated with the building of a compartmentalized site. In each case, the boundaries are permeable to only certain ions and molecules that are required for the mineralization process.

Vesicles, cells and macromolecular frameworks are assembled into enclosed semipermeable structures prior to biomineralization.

5.1.1 Phospholipid vesicles

As we discussed in Chapter 3, Section 3.3.1, vesicles are fluid-filled compartments enclosed by a continuous lipid bilayer (see Fig. 3.7). Their spontaneous self-assembly in aqueous solutions arises from the balancing of hydrophobic and hydrophilic interactions that exist for amphiphilic molecules dispersed in a polar solvent. In the absence of any external scaffolds, vesicles generally adopt a spherical morphology because this gives the minimum total surface energy for a given volume.

Phospholipids—see Fig. 5.1 for some examples—are key membrane constituents of biological vesicles. They readily form self-sealing biomolecular sheets, although packing of the molecules in the bilayer tends to be quite loose so there is usually a high degree of lateral fluidity involving the sideways shuffling of molecules around the inner or outer surfaces of the membrane (Fig. 5.2). Movement across the membrane—referred to as 'flip-flop'—is generally more difficult as it takes significant energy to move the polar or charged headgroup through the bilayer structure. Proteins are also present in the vesicle membrane, some of which span the entire 4 to 5 nm thickness of the bilayer. As we shall discuss in Section 5.3, these play an important role in the transport of ions and molecules into and out of the fluid-filled compartment.

Diphosphatidyl glycerol (cardiolipin)

Phosphatidyl ethanolamine

Phosphatidyl inositol

Phosphatidyl choline

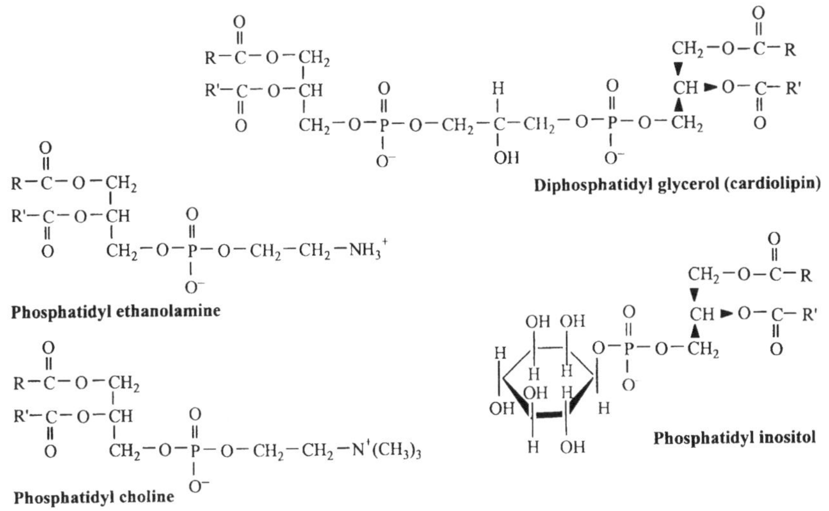

Fig. 5.1 Phospholipids (R and R′ are long-chain moieties).

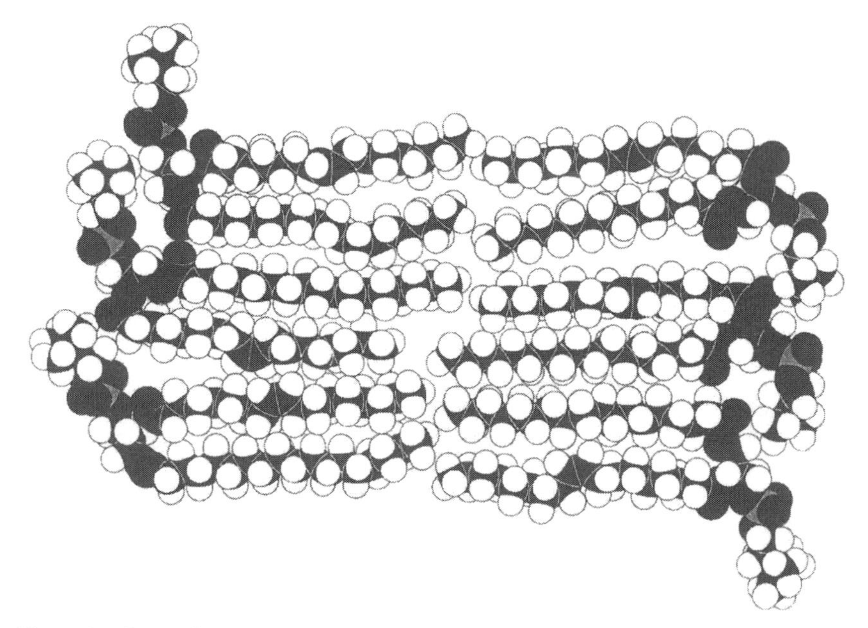

Fig. 5.2 Space-filling model of a fluid phospholipid bilayer membrane.

In general, phospholipid vesicles are ubiquitous in the cytoplasm of *eukaryotic cells* (cells containing a nucleus and nuclear membrane) where they are involved in a wide range of transport, storage and secretory functions. They are assembled with well-defined structures and membrane compositions, and originate from an organelle called the *Golgi complex* (Fig. 5.3). The Golgi complex is a stack of flattened membranous sacs called cisternae that act as a sorting centre in the targeting of proteins to various sites in the cell. The organelle is effectively a postal depot for proteins. Incoming proteins are 'written' in an adjacent organelle called the *endoplasmic reticulum* (ER) and then delivered inside vesicle envelopes to the *cis* face of the Golgi stack (G), as shown in Fig. 5.4. The vesicles (V) are stamped with a bristle-like coat that allows them to be recognized and unpackaged. Then the proteins are transferred through the stack, sorted and resealed in new vesicles and sent out to various destinations from the concave *trans* face. Although some of the exported vesicles (V*) remain fixed in specific locations within the cell, or fuse with the cell membrane, others are free to roam around usually by moving along protein filaments that criss-cross the cell interior. This trafficking means that the vesicles can undergo a variety of specific transactions with the cell walls or other vesicles where they release or encapsulate ions and molecules by membrane rupture, fusion and re-formation.

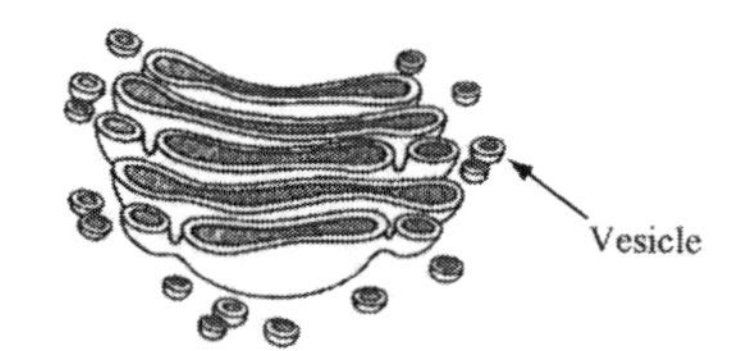

Fig. 5.3 Golgi complex and surrounding vesicles drawn in cross-section.

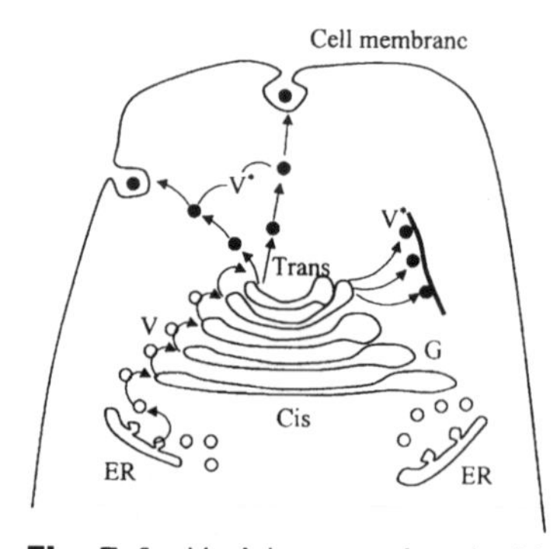

Fig. 5.4 Vesicles associated with the Golgi complex. ER, endoplasmic reticulum; G, Golgi complex; V, imported vesicles; V*, exported vesicles.

There is much experimental evidence that phospholipid vesicles are well established structurally in biomineralization prior to inorganic precipitation. For example, spherical vesicles are known to form in advance of iron oxide (magnetite, Fe_3O_4) nucleation in *magnetotactic bacteria* (Fig. 5.5A). The iron oxide crystals then nucleate on or close to the inside surface of the phospholipid membrane (Fig. 5.5B). Significantly, the formation of the magnetite crystals does not by itself result in magnetotaxis because a cellular compass

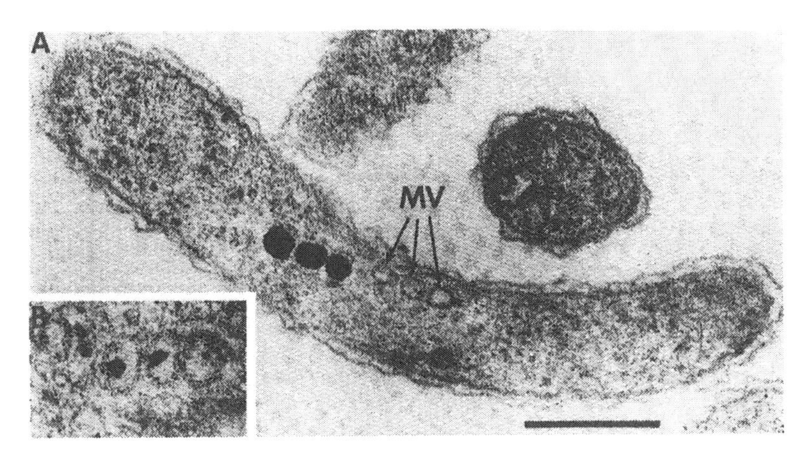

Fig. 5.5 Section through a magnetotactic bacterial cell showing: (A) three mature magnetite crystals and three empty magnetosome vesicles (MV); (B) vesicles containing immature magnetite particles. Scale bar, 250 nm.

needle is required. If the vesicles are disorganized in the cell then the magnetic fields arising from their associated crystals could cancel out with the result that there is no permanent dipole moment in the cell. The bacteria solve this problem by organizing the crystals into relatively rigid chains by arranging the vesicles in a linear array closely associated with the inner surface of the cell wall. Formation of the magnetite crystals then proceeds through a sequence of events in which the vesicle membrane is of paramount importance. As shown in Fig. 5.6, these include:

- uptake of Fe^{III} ions from the environment
- reduction of Fe^{III} to Fe^{II} ions during transport across the cell membrane
- transport of Fe^{II} ions to and across the vesicle membrane (the *magnetosome* membrane)
- precipitation of amorphous hydrated Fe^{III} oxide within the vesicle
- transformation of the amorphous phase to magnetite by surface reactions involving mixed valence intermediates (see Chapter 4, Section 4.10.3).

The single crystal nature of the bacterial magnetite crystals implies that the nucleation of magnetite from the amorphous Fe^{III} oxide precursor occurs at one primary site within the vesicle. For example, if crystal nuclei do initi-

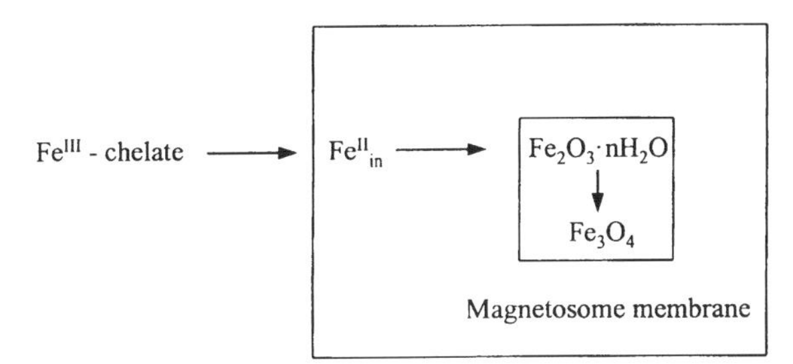

Fig. 5.6 Ion transport and redox processes associated with the formation of bacterial magnetite.

ate at other sites, they must rapidly dissolve and reprecipitate at the primary site. It seems most probable that the surrounding vesicle membrane plays a crucial role not only in imposing a spatial constraint on this process but also in controlling nucleation. In this regard, biochemical analysis of the membrane surrounding the bacterial magnetite crystals shows that the overall composition is similar to conventional membranes except for two specific proteins, which are presumably inserted into the lipid bilayer (see further reading). One possibility is that these protein molecules are locally clustered so that they promote nucleation at one specific site along the membrane.

The high structural quality of the bacterial magnetite crystals (see Fig. 3.3 in Chapter 3) suggests that the chemical composition within the vesicle must be finely controlled. In particular, as the phase transformation process depends on the reduction of a third of the Fe^{III} ions and releases protons (see Chapter 4, Section 4.10.3), there has to be very good control over both the redox and pH if the reaction is to proceed to completion. Moreover, a simple change in the concentration of anions such as Cl^-, SO_4^{2-} and $H_2PO_4^-$ is known to have a marked influence on magnetite crystallization in the laboratory, so the concentration of extraneous ions within the vesicle must also be highly regulated. For more details, see further reading.

Phospholipid vesicles exported from the Golgi complex are commonly associated with the spatial and chemical control of biomineralization processes. In magnetotactic bacteria, vesicles house a series of complex chemical processes that result in well-ordered single crystals of magnetite.

5.1.2 Protein vesicles—ferritin

The iron storage protein *ferritin* (see Chapter 2, Section 2.6.2) is an unusual example of a fluid-filled vesicle that is constructed from polypeptide building blocks. The protein consists of a spherical polypeptide cage, 8 nm in internal diameter and 2 nm in thickness, which arises from the spontaneous self-assembly of 24 polypeptide subunits (Fig. 5.7). In mammalian ferritin, there are two different subunits, referred to as heavy (H) and light (L), which have molecular masses of around 21 000 and 20 000, respectively. The relative proportion of the H and L subunits depends on the biological source (rat, human, horse, etc.) and varies between different tissues. For example, horse spleen ferritin contains 10 per cent H-chain and 90 per cent L-chain subunits whereas ferritin extracted from equine heart is a mixture of 90 per cent H-chain and 10 per cent L-chain. Nevertheless, there is usually a high degree of identity, often with 80 to 90 per cent of the residues conserved, in the amino acid sequences between the various H-chains or L-subunits.

Both subunits have very similar secondary and tertiary polypeptide structures based on a bundle of four long α-helices—labelled as A, B, C and D in Fig. 5.7B—a short helix (E) and a long loop (L) of 17 amino acid residues. The bundle is composed of two pairs of antiparallel helices (AB, CD) and the loop connects helix B to helix C in a way that places A and B in antiparallel orientation (Fig. 5.8). Helix E is at the carboxy end (C-terminus) of the amino acid chain (amino acids are numbered from the uncoupled amino end (N-

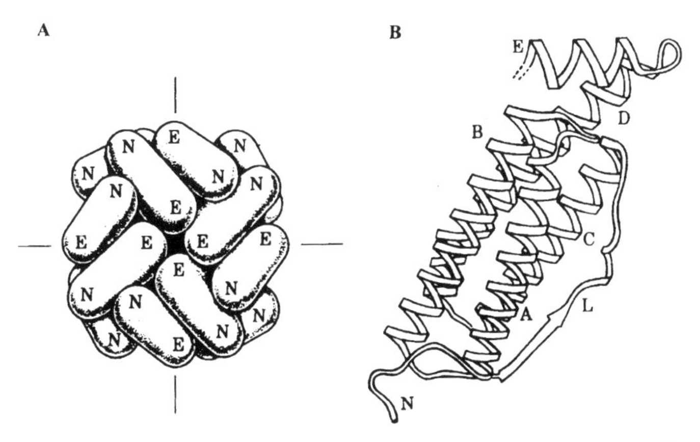

Fig. 5.7 Ferritin. (A) Protein shell and arrangement of subunits: N, amino-terminus; E, carboxy-terminus of polypeptide chain. (B) Single subunit showing bundle of four α-helical domains (A–D), loop region (L), and small helix (E) of the polypeptide chain.

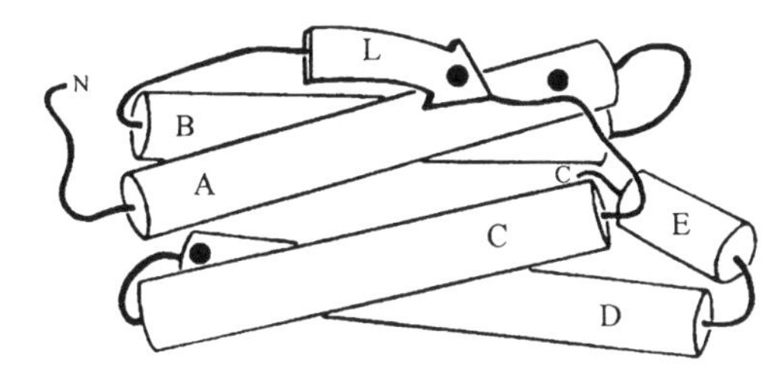

Fig. 5.8 Simplified representation of ferritin subunit structure in which the α-helices are shown as cylinders.

terminus) of a protein) and aligned at approximately 60° to the bundle axis. Differences in the H- and L-chain subunits usually arise from changes in the number of amino acids present in the short stretches that occur before and after helices A and E, respectively, rather than variations in the helical and loop secondary structures.

Self-assembly of the protein shell is in many respects analogous to the formation of vesicles from assemblies of phospholipid molecules. The supramolecular interactions involve hydrophobic and hydrophilic forces that together drive the system into an organized structure. In the case of ferritin, this results in the formation of *subunit dimers* with a well-defined molecular structure based on the specific interdigitation of the external surfaces of both the loop section and helix A of an antiparallel pair of polypeptide chains. However, this arrangement is only metastable because the hydrophobic helix E of both subunits—shown schematically in Fig. 5.9A as the shaded regions at the ends of the curved rods—is exposed to the polar solvent. As a consequence, the subunit dimers continue to aggregate through a series of stages involving additional intersubunit interactions, such as hydrogen bonding, salt bridges and hydrophobic forces, until 12 dimers are incorporated into a closely packed hollow shell with all the E helices buried (Fig. 5.9B–D).

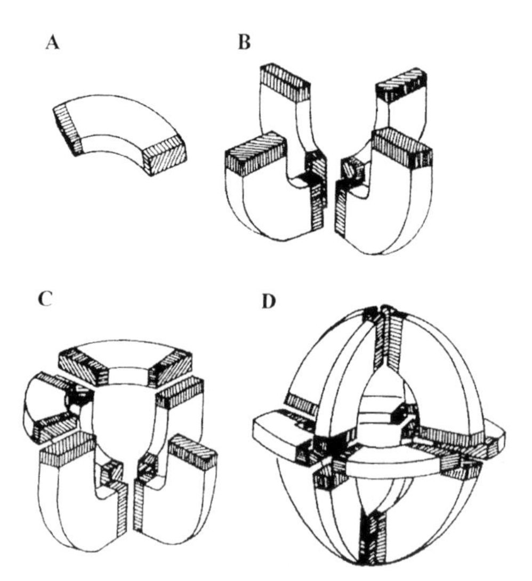

Fig. 5.9 Self-assembly of ferritin shell from subunit dimers. See text for details.

One of the most important consequences of this amazing process of self-assembly is the perforation of the protein shell by small channels that lie on the three- and fourfold symmetry axes located at the junctions of the subunit dimers. These channels are about 0.3 nm wide and there are eight hydrophilic and six hydrophobic pores along the three- and fourfold axes, respectively. The latter are surrounded by four E helices, as shown in cross-section in Fig. 5.10 (see also Fig. 5.7). By comparison, the hydrophilic channels arise from the juxtaposition of the D helix and the C–D turn. In both cases, the channels are lined by specific amino acids. For example, in mammalian ferritins three aspartate (Asp) and three glutamate (Glu) residues are located along the wall of the hydrophilic pores, whereas 12 leucine (Leu) residues protrude into the hydrophobic channel (Fig. 5.11). A molecular scale view down the fourfold hydrophobic channel is shown in Fig. 5.12.

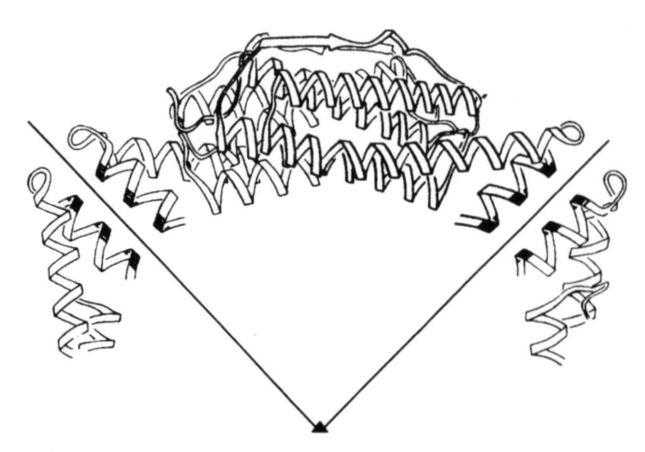

Fig. 5.10 Cross-section of ferritin shell showing hydrophobic channels and associated E helices of the subunit dimers.

```
  H  H  O           H  H  O           H  H  O
  |  |  ||          |  |  ||          |  |  ||
-N- C - C-        -N- C - C-        -N- C - C-
    |                 |                 |
    CH2               CH2               CH2
    |                 |                 |
    C                 CH2               CH
  //  \               |                /  \
 O     O-             C             H3C    CH3
                    //  \
                   O     O-

    Asp               Glu               Leu
```

Fig. 5.11 Amino acid side chains.

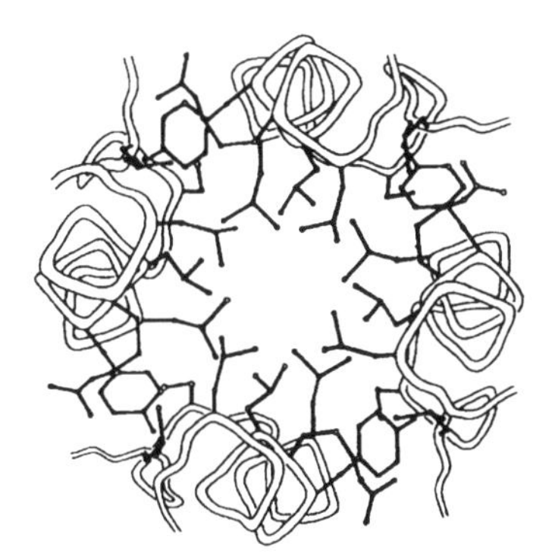

Fig. 5.12 View down a fourfold channel lined with 12 leucine residues.

The molecular channels in the ferritin shell are required for the transport of Fe^{II} ions and other species into and out of the aqueous cavity for iron oxide (ferrihydrite) mineralization. Iron transport is most likely to occur via the hydrophilic threefold channels, which have been identified as sites for metal ion binding. Overall, the formation of the mineral core is a complex process with several sites in the protein playing key roles in Fe^{II} oxidation and Fe^{III} hydrolysis and condensation. Further details on the nucleation process are described in Chapter 6, Section 6.7.3.

Ferritin is a protein vesicle consisting of a polypeptide shell with well-defined molecular channels. The shell is self-assembled from polypeptide subunit dimers that spontaneously organize to produce an aqueous-filled cage with high symmetry and specificity for iron oxide (ferrihydrite) biomineralization.

5.1.3 Cellular assemblies

Another method of sealing off biological space for mineralization involves groups of cells getting together. For example, one theory of bone growth proposes that bone-forming cells—*osteoblasts*—collectively assemble to produce a fluid-filled compartment that is separated from the blood and contains the mineralized tissue (Fig. 5.13). The layer of cells is tightly packed so that the chemistry of the inner compartment (IC) is regulated by controlled diffusion of ions and molecules from the osteoblasts (OB). On one side of the membrane, the osteoblasts are in direct contact with bone cells—*osteocytes* (OC)—that form an interconnected network inside the bone, and on the outer side they are exposed to the extracellular fluid (EF), which is in equilibrium with the blood. Thus, there is a direct line of communication between the external and internal environments that enables the bone to grow, remodel or remain in a steady state depending on the hormonal signals arriving in the blood.

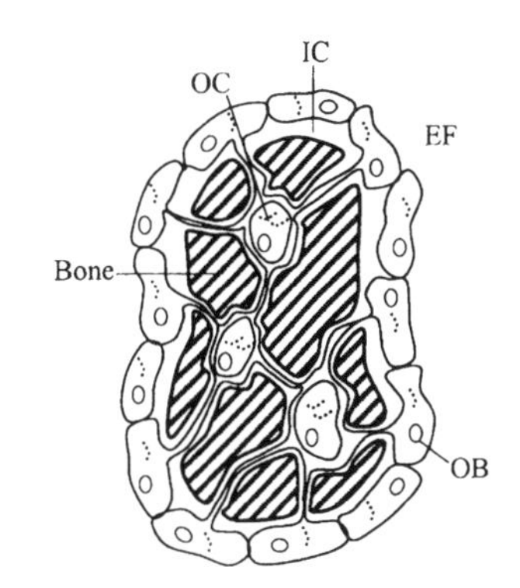

Fig. 5.13 Cellular assemblies in boundary-organized bone formation. IC, inner compartment; OB, osteoblasts; OC, osteocytes; EF, extracellular fluid.

Alternatively, cells can group together in association with a solid substrate such as a polymerized layer of macromolecules or a pre-existing mineral surface (Fig. 5.14). This occurs, for example, in the outer and inner shell layers of molluscs, respectively (see Fig. 2.2 in Chapter 2). The outer *prismatic layer* consists of long calcite crystals that develop in the space between a closely packed sheet of epithelial cells and a highly insoluble protein layer called the *periostracum*, which covers the external surface of the shell. In contrast, the aragonite crystals of the inner layer are deposited after the

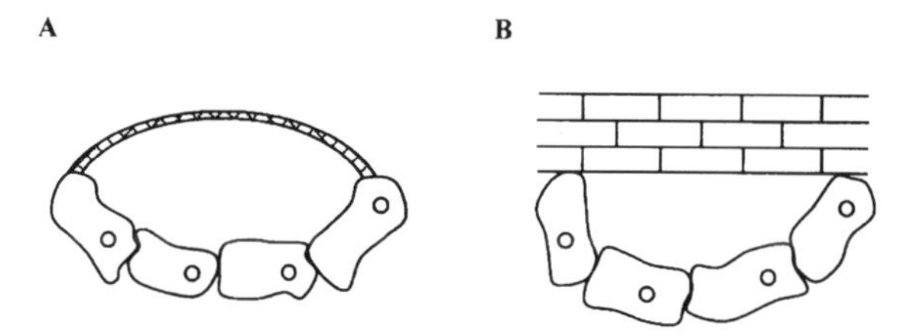

Fig. 5.14 Space delineation using groups of cells in association with: (A) an impervious polymerized organic sheet; (B) an existing mineralized structure.

calcite layer so the mineralization site is confined to the space between the cell layer and the growth front of the mineralized structure.

Assemblies of cells, often in association with impervious substrates, are used to seal off mineralization sites and control the chemistry of the internalized fluid-filled environment.

5.1.4 Macromolecular frameworks

Cells are often tens of micrometres in size so the spaces produced by their assembly are large when compared with those delineated by vesicles. This has the potential disadvantage that the growth of the mineral crystals could easily get out of biological control. To prevent this, organisms often partition the mineralization space formed by cellular groupings into smaller enclosures through the use of semipermeable organic matrices with open framework structures. The growth of the mineral phase then becomes contained at a more local level where it can be spatially organized.

In the *avian eggshell*, for example, calcite crystals grow in a spongy organic matrix that is secreted between a layer of epithelial cells and an insoluble fibrous membrane of keratin-like proteins (the stuff of hair and nails) and carbohydrates. As shown in Fig. 5.15, the crystals nucleate at specific sites on the membrane (see also Fig. 3.8 in Chapter 3) and then develop as polycrystalline outgrowths (*spherulites*) that extend along their crystallo-

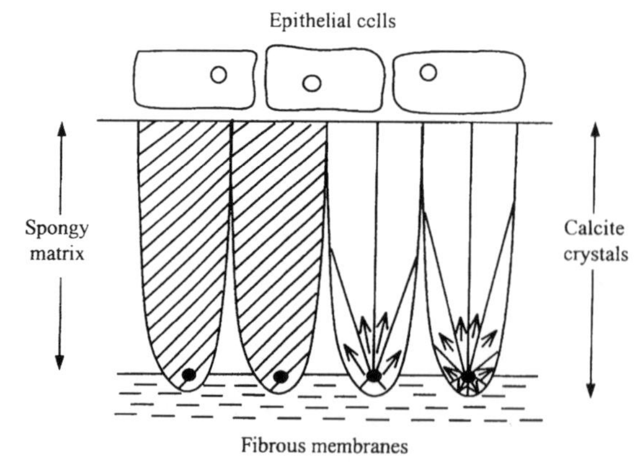

Fig. 5.15 Eggshell formation. Arrows indicate directions of calcite *c* axes in the polycrystalline outgrowths.

graphic *c* axes in all directions into the spongy matrix. Although it subdivides the mineralization space and contributes to the mechanical strength of the eggshell, the spongy matrix has very little influence on the growth of the crystals, which continue unabated until neighbouring spherulites come into contact laterally. With time, and as the shell thickens, this results in an apparent crystallographic alignment between the nucleation centres because fewer and fewer crystals are being extended outwardly from the surface of the membrane. Clearly, the thickness at which this occurs depends on how far apart the original nucleation centres are, so that shells with very closely spaced sites have outgrowths that quickly become crystallographically aligned. These relationships have been studied by X-ray diffraction (see further reading).

Other biological systems use macromolecular frameworks to provide a significant degree of spatial control over the mineralization process. For example, *limpet teeth* contain arrays of needle-like iron oxide crystals with the goethite (α-FeOOH) structure (Fig. 5.16). The crystals are aligned parallel to each other by a series of regularly spaced organic filaments and tubules that restrict the crystal thickness within the range 30 to 50 nm, and direct the growth preferentially along the crystallographic *c* axis. Because the matrix is made of *chitin* (a polysaccharide) it is initially soft and pliable, and can be twisted and interwoven into a complex arrangement of enclosed spaces that subsequently become filled with the needle-shaped crystallites to produce a strong and tough reinforced fibre composite. See further reading for more details.

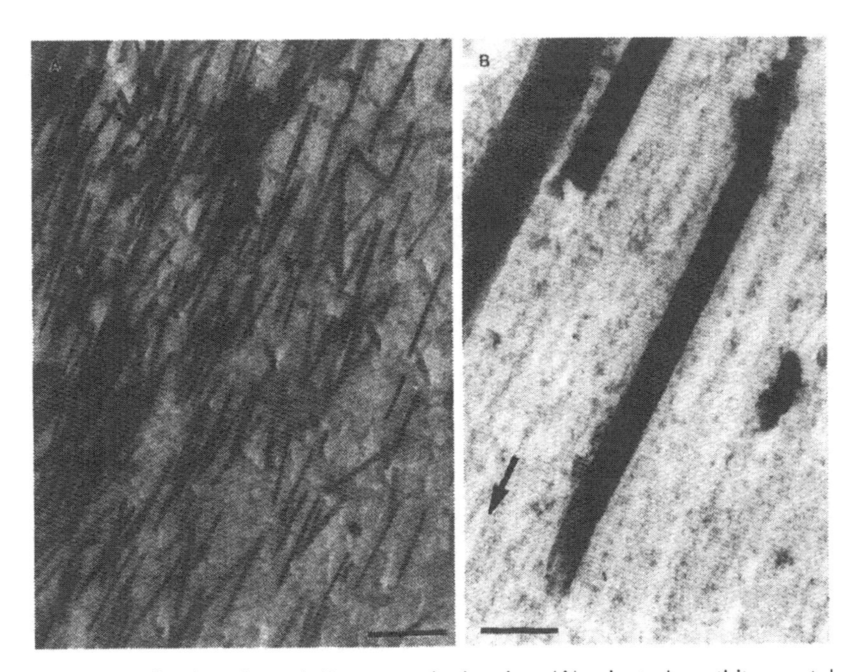

Fig. 5.16 Section through limpet teeth showing: (A) oriented goethite crystals (scale bar, 600 nm); (B) single crystal and associated chitin filaments (arrow) (scale bar, 100 nm).

An assemblage of insoluble proteins, sometimes in association with chitin fibres, is used to produce one of the most remarkable examples of spatial control in biomineralization—the aragonitic *nacreous layer* of seashells. As we mentioned in Section 5.1.3, nacre biomineralization is confined to the space between the epithelial cell layer and the mineral growth front. But as shown in Fig. 5.17, this space is further partitioned by a series of organic sheets (S) that lie approximately parallel to the layer of epithelial cells (EC). The sheets are secreted sequentially at the mineralization front so that with time the shell increases in thickness by inward growth as the spaces become filled with tablet-shaped aragonite crystals. Each crystal is surrounded by an organic envelope (E) and sandwiched between two organic sheets that are regularly spaced at a distance of approximately 0.5 μm, and this gives rise to the highly uniform thickness of the tablets. In contrast, growth in the plane of the sheets is less restricted and the crystals develop until they come into contact with other aragonite tablets in the same layer to produce a crazy-paving structure. As the layer begins to fill up with crystals, a new sheet is deposited and the process is reinitiated.

In *gastropod* shells—such as the single coiled shell of a snail—the timing of this episodic process results in a remarkable 'stack-of-coins' structure at the growth surface in which the crystals between successive sheets appear to nucleate close to the centre of pre-existing tablets located in the layer below (Fig. 5.17A). In fact, the alignment is facilitated by the growth of crystals through nanopores in the organic sheet such that small mineral bridges traverse the organic matrix and connect the crystals within each individual stack. In *bivalve* shells, however, each successive layer of crystals is offset, and the lateral growth of the aragonite tablets occurs to a much greater extent before the next sheet is added. The resulting structure is then more akin to a 'brick wall' (with a protein–chitin mortar) than a 'stack of coins' of diminishing size (Fig. 5.17B). For more details on shell mineralization, see Chapter 6, Section 6.5, and further reading.

The partitioning of sealed-off environments by insoluble macromolecular frameworks is common in extracellular biomineralization. The matrix influences the mechanical properties and in many cases controls the spatial organization of large numbers of mineral crystals. Shell nacre is spatially patterned by the episodic deposition of a protein–chitin matrix, and variations in this process account for differences between the architecture of gastropod and bivalve shells.

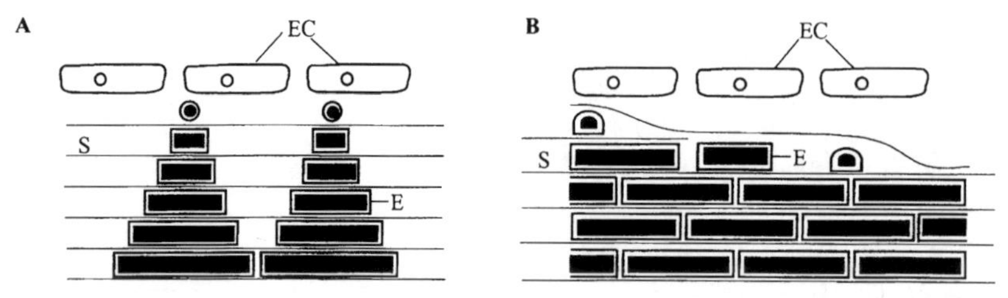

Fig. 5.17 Nacre formation in (A) gastropod and (B) bivalve seashells. Aragonite crystals are shown in black. EC, epithelial cells; S, sheets; E, envelopes.

5.2 Supersaturation control within spatial boundaries

The delineation of biological environments as described in Section 5.1 provides not only spatial control on the nano- to macroscopic scale but also a means for regulating the solution chemistry within fluid-filled vesicle compartments and macromolecular frameworks. In this section we discuss a general model for controlling the supersaturation level of these enclosed spaces.

In general, the level of supersaturation within intracellular vesicles and macromolecular frameworks is controlled by *membrane transport*. In the former, the barrier to ion flow is the lipid bilayer of width 4 to 5 nm, while supersaturation levels in extracellular environments are controlled by ion transport across neighbouring cell membranes, as well as passive diffusion through the organic matrix. In both cases, we can consider two fundamental types of mechanisms that either directly or indirectly increase the supersaturation within the confined space (Fig. 5.18).

The direct mechanisms include:

- *ion pumping*—both cations (M^{n+}) and anions (X^-) can be selectively transported by a number of membrane-based processes to increase the ionic concentration (see Section 5.3 for further details). Some of these are facilitated by redox changes at the membrane surface, for example Fe^{III} to Fe^{II}.
- *ion complexation*—the binding of transported cations with ligands such as citrate or pyrophosphate produces complexes (MC) that lower the supersaturation, but this is re-established by the controlled release of the cations by subsequent decomplexation.
- *enzymic regulation*—removal of H_2CO_3 by the enzyme, *carbonic anhydrase*, can be coupled to calcification processes, according to the equilibrium

$$Ca^{2+} + 2HCO_3^- \rightleftharpoons CaCO_3 + H_2CO_3$$

  Similarly, the enzyme, *alkaline phosphatase*, releases HPO_4^{2-} ions at the sites of calcium phosphate mineralization, for example inside vesicles present in the extracellular matrix of cartilage.

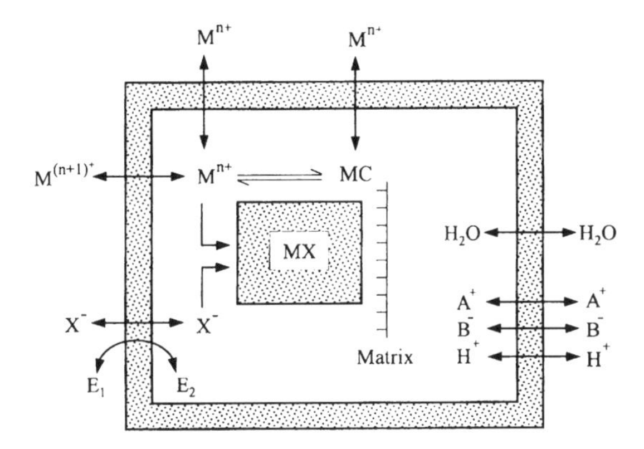

Fig. 5.18 Control of localized supersaturation. See text for details.

Alternatively, there are several indirect mechanisms to increase supersaturation within the enclosed compartment:

- *ionic strength*—increases in activity products can be induced by changes in the background electrolytes arising from the selective transport of non-mineralizing ions such as Na^+ and Cl^-.
- *water extrusion*—increases in activity products and concentrations occur from water loss; for example, silica deposition is induced in the leaves of plants by transpiration.
- *proton pumping*—changes in pH by transport of H^+ ions or localized changes in cellular metabolism influence the acid–base equilibria of oxyanions, such as HCO_3^-/CO_3^{2-} and HPO_4^{2-}/PO_4^{3-}, as well as the hydrolysis of metal ions like Fe^{III}. For example, magnetite (Fe_3O_4) formation results in the release of H^+ ions that must be removed from the mineralization site if the following reaction is to go to completion:

$$Fe_2O_3 \cdot nH_2O + FeOH^+ \rightarrow Fe_3O_4 + H^+ + nH_2O$$

In the next section we describe some of the general features of ion transport through biological membranes.

The control of supersaturation levels in biomineralization involves the chemical regulation of localized environments through ion transport, complexation–decomplexation, enzymatic regulation, and modifications in ionic strength, redox, pH and water content.

5.3 Ion transport

Since the dawn of life, cells have adopted a siege mentality with regard to transactions with the external environment. As *fortress membranous*, they have been able to shield themselves from a hostile world, but at considerable energy cost. The main problem is that the high levels of organization required for life are only maintained if the system remains open with respect to the environment. Otherwise, the overall entropy production is negative, and this violates the second law of thermodynamics! Phospholipid bilayers cannot therefore be totally impenetrable because certain ions and molecules (and energy) must be continually exchanged between the intra- and extracellular environments. So the evolution of the cell membrane is contingent on the presence of various different types of gatekeepers associated with the bilayer structure. These are usually in the form of multi-subunit complex proteins that span the phospholipid membrane and act as transport channels for specific types of ions and molecules.

There are various types of passport control governing the immigration and emigration of ions and molecules according to their chemical credentials. The main types involve either the metabolic *pumping* of specific ions, often against an electrochemical gradient, or a two-way exchange (*antiport* mechanism) that is driven by electrochemical gradients (Fig. 5.19). The metabolic pump is based on *active* transport because the change in free energy accompanying ion transport is positive and an energy source is therefore required.

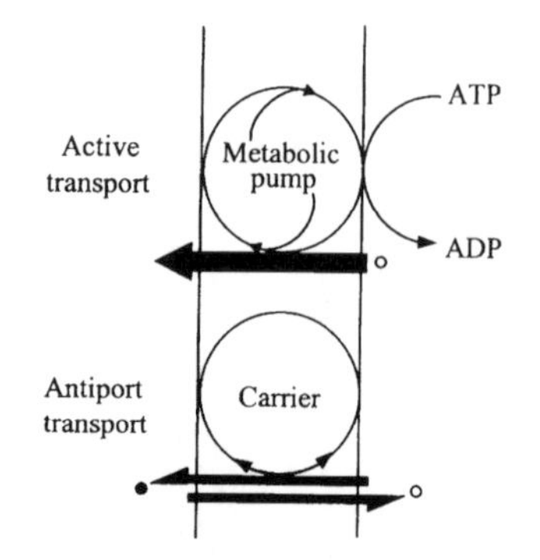

Fig. 5.19 Mechanisms of ion transport across the cell membrane.

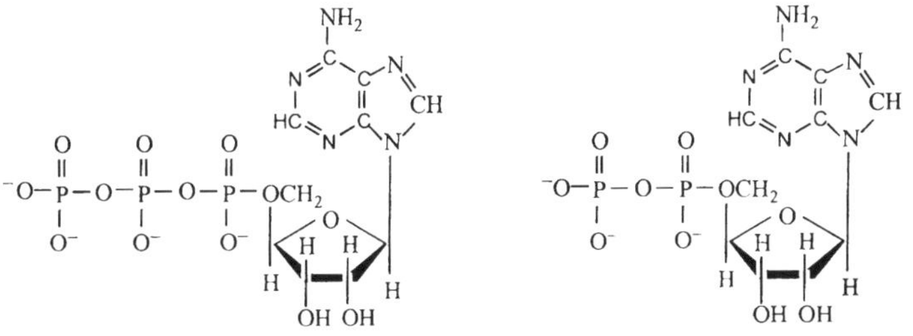

Adenosine triphosphate
(ATP)

Adenosine diphosphate
(ADP)

Fig. 5.20 Energy-rich nucleotides.

Many of the common ion pumps use the hydrolysis of adenosine triphosphate (ATP) to adenosine diphosphate (ADP) (Fig. 5.20) as a supply of localized energy to transport Na^+ ions out of the cell and import K^+ ions. Similar *ion-transporting ATPases* are involved with the pumping of H^+ and Ca^{2+} ions out of the cell or into intracellular vesicles (Fig. 5.21).

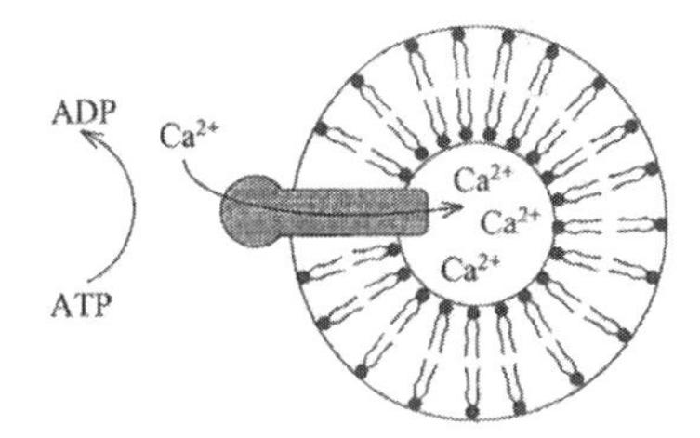

Fig. 5.21 Ca^{2+}-ATPase pump in vesicle membrane.

The forced expulsion of ions through metabolic pumps involves subtle structural changes in the helical domains of multi-subunit transmembrane proteins. These are associated with ATP binding and phosphorylation of an aspartic acid residue on the polypeptide chain exposed on the inner surface of the cell membrane (Fig. 5.22). This results in the formation of a transient channel and the translocation of the ions. The channel then closes unless reactivated.

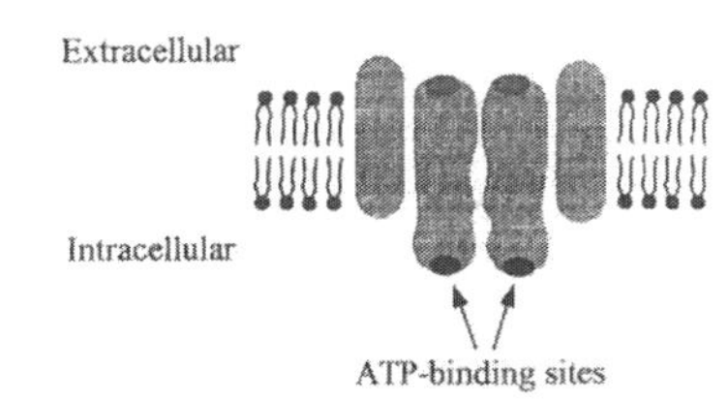

Fig. 5.22 Na^+–K^+ pump with subunits, α and β, and intracellular ATP binding sites.

Whereas the ATP-driven pumping of ions through a cell or vesicle membrane often involves the translocation of a single type of ion, the antiport system always requires a counter exchange. For example, Ca^{2+} ions are transported out of the cell against a chemical gradient by a counterflow of H^+ or Na^+ ions that move down a gradient in the opposite direction. As there is approximately a 10^4 M difference in the Ca^{2+} ion concentration between the intra- and extracellular fluids—the former is around 10^{-7} M—then a lot of energy is required to move the Ca^{2+} ions against such a large electrochemical gradient. In fact, in many systems a network of Ca^{2+}-ATPase pumps and Ca^{2+}/Na^+ antiporters work in parallel to maintain the low intracellular calcium level.

Anions are often transported across cell membranes through preformed hydrophilic channels that traverse the phospholipid bilayer. Transmembrane proteins snake back and forth across the membrane to produce a cylinder of closely knit α-helical bundles that enclose a central pore (Fig. 5.23). The channel is more or less open at all times so that anions move through by passive transport down an electrochemical gradient. This often involves an antiport exchange; for example, HCO_3^- ions inside blood cells are exchanged one-for-one with Cl^- anions such that the electrochemical potential of the cell membrane remains unchanged.

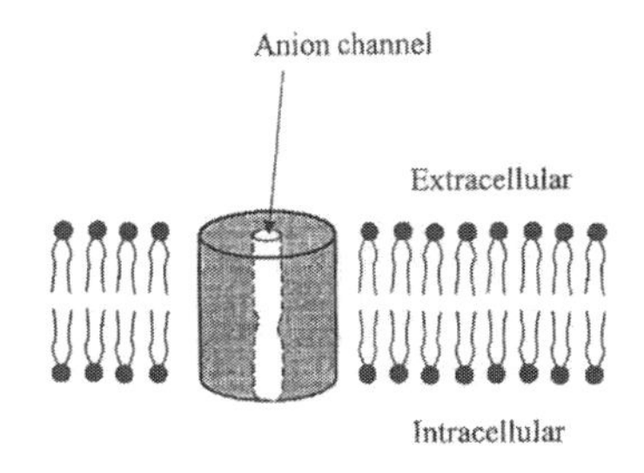

Fig. 5.23 Anion channel in cell wall membrane.

Having described the main types of ion transport systems, we now discuss in the next section the role of ion fluxes in calcification processes.

Ion flow across phospholipid membranes is an important aspect of boundary-organized biomineralization. Cations are often pumped against a concentra-

tion gradient by ion-transporting ATPases, or exchanged for other cations that are flowing down a gradient in the opposite direction. Anions are exchanged through channel-forming transmembrane proteins by passive transport.

5.4 Ion fluxes in calcification

The forced export of Ca^{2+} ions from the internal fluid of a cell is a common feature of life. It is simply a matter of survival because Ca^{2+} ions bind very strongly to the phosphate-rich compounds of energy metabolism, such as ATP, and inhibit their activity. Likewise, many proteins become cross-linked and precipitate in the presence of Ca^{2+} ions. For these reasons, the level of calcium inside the cell is usually of the order of 10^{-7} M compared with 10^{-3} M in the extracellular fluids. Cells are therefore continually expelling Ca^{2+} ions back across the cell wall or annexing them in intracellular vesicles. In this regard, vesicles are like little drops of extracellular fluid adrift in a cytoplasmic ocean.

It is therefore not too surprising that the shuttling of cellular calcium is closely linked with cell wall and vesicle-centred calcification processes because the immobilization of the high levels of Ca^{2+} ions in the form of a non-toxic mineral phase clearly helps to keep things under control. Moreover, provided that the inorganic deposit does not interfere with the biochemical machinery, then a lot of calcium, along with carbonate and phosphate, can be locked away safely and kept in mineral storage for future use if needs be.

Life is dependent on the continual transport of Ca^{2+} ions out of the cell and into intracellular vesicles. The associated fluxes of calcium are often linked to biomineralization processes and probably to the early evolution of calcified structures in biology.

5.4.1 Calcification in green algae

Algae are tiny single-celled plants that often live in colonies in marine and freshwater environments. A wide variety of these organisms deposit calcium carbonate in association with the transport of ions across organic membranes (see further reading). In certain species of freshwater green algae called *Chara*, calcification occurs adventitiously on the external surface of the cell wall, but paradoxically this is linked to the *uptake* of HCO_3^- into the cells. How this works is shown in Fig. 5.24. The key point to note is that there is a spatial inhomogeneity in the operation of cell wall pumps for H^+ and OH^- ions. Together, protons and HCO_3^- ions in the extracellular medium are transported into the cell through an ion-channel referred to as a *symport*. However, whereas the HCO_3^- ions are converted by carbonic anhydrase within the cell into CO_2 and OH^- ions, the protons are pumped back to the extracellular medium by a transmembrane ATPase, where they partner other HCO_3^- ions on a journey back into the cell. This recycling of the H^+ ions takes place in close vicinity so a band of acidic pH is maintained just outside the cell wall. In contrast, the OH^- ions are transported out of the cell by a different trans-

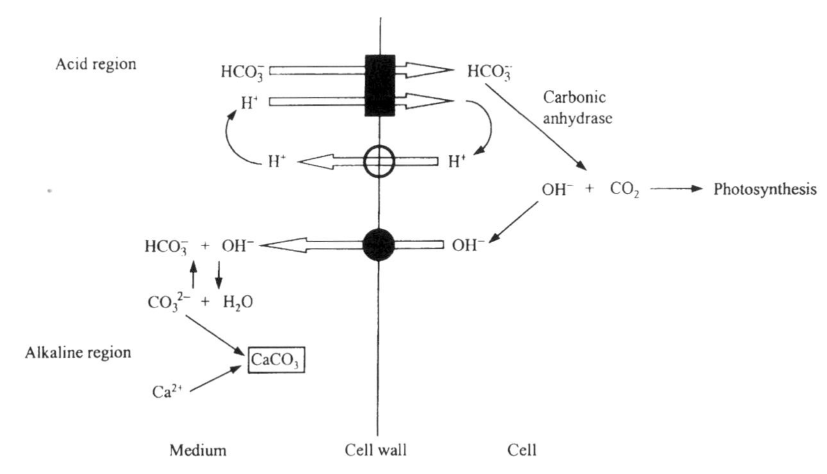

Fig. 5.24 Ion fluxes involved with calcite deposition on the cell wall of *Chara*.

membrane system that is spatially separated from the H^+ pumps. Because the cells are sufficiently large, a localized band of high pH—often as high as 10.5—builds up further along the external surface of the cell wall. In this region, HCO_3^- ions in the external medium become deprotonated and a high concentration of CO_3^{2-} anions ensues. With sufficient levels of Ca^{2+} in the freshwater environment, this leads to supersaturation and the deposition of a discrete band of calcium carbonate (calcite) on the cell wall.

The CO_2 produced in the above mechanism is used for *photosynthesis*. This process occurs in the presence of sunlight within organelles called *chloroplasts* and involves the fixation of CO_2 as reduced carbon in organic molecules such as sugars and carbohydrates ($[CH_2O]$). This is often represented as:

$$CO_2 + H_2O \rightarrow [CH_2O] + O_2$$

Although in *Chara* this occurs in the absence of Ca^{2+}, in other types of green algae the removal of CO_2 from a boundary-organized site containing Ca^{2+} ions has a direct influence on $CaCO_3$ deposition. This arises by changes in the balance of the acid–base equilibria arising from the dissolution of CO_2 in water, which involve the following aqueous species:

$$CO_2 + H_2O \rightleftharpoons H_2CO_3$$
$$H_2CO_3 \rightleftharpoons H^+ + HCO_3^-$$
$$HCO_3^- \rightleftharpoons H^+ + CO_3^{2-}$$

Taken together, these equilibria can be represented by the interconversion

$$CO_2 + H_2O \rightleftharpoons 2H^+ + CO_3^{2-}$$

which indicates that the removal of CO_2 by photosynthesis drives the equilibrium to the left, consuming protons and increasing the solution pH. Although CO_3^{2-} ions would also be consumed, they are buffered by further deprotonation of HCO_3^- as the pH increases. The net effect is that the concentration of

CO_3^{2-} ions rises, which in the presence of sufficient amounts of Ca^{2+} ions increases the supersaturation with respect to $CaCO_3$ deposition, that is

$$CaCO_3 \rightleftharpoons Ca^{2+} + CO_3^{2-}$$

so that overall the relationship between photosynthesis and calcification can be written as

$$CaCO_3 + CO_2 + H_2O \rightleftharpoons Ca^{2+} + 2HCO_3^-$$

and the equilibrium lies to the left when CO_2 is removed by photosynthesis.

This raises a fundamental question concerning the relationship between ion fluxes, calcification and photosynthesis: is the biomineralization of calcium carbonate a spin-off from the general photosynthetic requirement to extract CO_2 from carbonated aqueous fluids? This question has fascinated biologists for many years and the answer seems to be 'maybe'! One example where there seems to be a good correlation between these processes is the formation of needle-like crystals of aragonite in the marine green alga *Halimeda*. This system involves the boundary-organized biomineralization of aragonite within the *intercellular* spaces that are sealed off from the adjacent cells and the surrounding seawater (Fig. 5.25). Ions can slowly diffuse into the compartment from the seawater so the composition of the space is very similar to the external medium. However, because the diffusion path into the intercellular space is so long—of the order of several hundreds of micrometres—then changes in the composition of the entrapped fluid are not immediately rectified. Uptake of CO_2 from the compartment by the bordering cells therefore results in a temporary increase in the pH sufficient to promote the deposition of $CaCO_3$ within the intercellular space.

Because photosynthesis occurs only in the presence of sunlight, it follows that the biomineralization process must also be stimulated by light and should

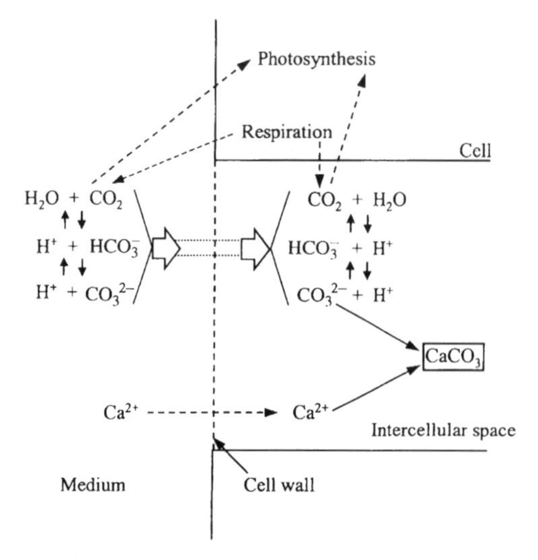

Fig. 5.25 Ion fluxes involved with the intercellular deposition of aragonite in *Halimeda*.

stop in the dark. Indeed, this correlation holds for *Halimeda* provided that the pH in the intercellular space is in the range of 6.6 to 8.5—any higher and the photosynthetic rate is inhibited. In fact, some of the aragonite crystals are partially dissolved in the dark because CO_2 diffuses back into the intercellular space from cellular respiration (effectively the converse of photosynthesis) and acidifies the environment.

Calcification in algae is often closely related to ion fluxes that directly or indirectly result in localized supersaturation. In the freshwater green alga Chara, calcite crystals are deposited adventitiously on the external surface of the cell due to spatially separated ion fluxes. Photosynthesis and aragonite mineralization are coupled in the marine green alga Halimeda through the removal of CO_2 and associated increase in the pH of localized intercellular spaces containing Ca^{2+} ions.

5.4.2 Coccolith calcification

Unlike their colonial green algal counterparts, phytoplanktonic algae, known as *coccolithophores*, are individualistic, swimming around in huge numbers in the surface layers of the world's oceans where they fix tons and tons of CO_2. Besides their environmental importance, they are distinguished by a remarkable propensity to produce wonderfully sculpted calcite scales, called *coccoliths*, during certain stages of their life cycle (see Fig. 2.3 in Chapter 2). As shown in Fig. 5.26, coccoliths (C) are formed in intracellular vesicles derived from the Golgi complex (G). The vesicles are flattened like pancakes and with time move away from the Golgi stack towards the cell boundary. During this migration other smaller vesicles attach themselves and inject polysaccharides and other components for the assembly of a single organic scale (OS) or base-plate within each flattened vesicle. The scales and their associated vesicles then continue together on their way to the cell membrane (CM) where they are assembled around the organism. In some of the Golgi-derived vesicles, however, crystals of calcite are nucleated around the rim of the organic base plate so that complex calcified scales are also produced during the journey to the outer membrane. These coccolith vesicles (CV) are then assembled around the cell wall as a *coccosphere* (CS). Further details on the construction and assembly of coccoliths are described in Chapter 8, Section 8.4.1.

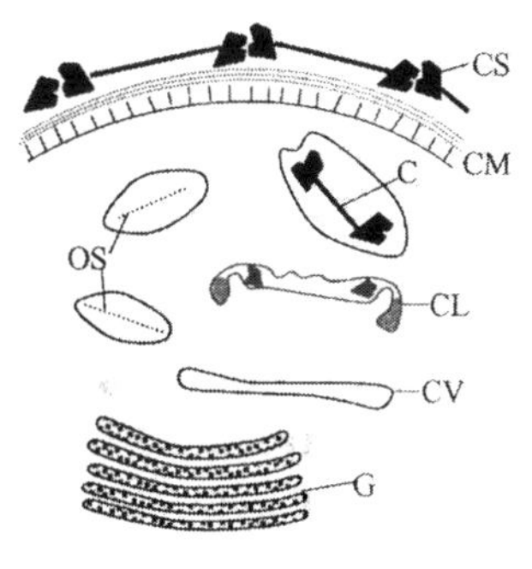

Fig. 5.26 Coccolith calcification in intracellular vesicles. G, Golgi complex; OS, organic scale; CM, cell membrane; C, coccolith; CV, coccolith vesicles; CS, coccosphere; CL, coccolithosomes.

Coccolith calcification in the coccolithophore *Emiliania huxleyi* involves the uptake of Ca^{2+}, HCO_3^- and CO_2 from the environment, diffusion through the cytoplasm and transport across the lipid bilayer of the Golgi-derived flattened vesicle (Fig. 5.27). See further reading for more details. Sulfate ions are also required because the major organic component of the calcite coccoliths consists of *sulfated polysaccharides*. Overall, the uptake of these ions and molecules is dependent on the presence of light, although to a different degree in each case. Calcium ions are transported across the cell wall down a large electrochemical gradient probably involving a uniporter. Then they diffuse directly to the coccolith membrane or are accumulated as Ca^{2+}-polysaccharide-rich nanoparticles of diameter 25 nm, referred to as *coccolithosomes*. These structures reside inside small special vesicles that bud off

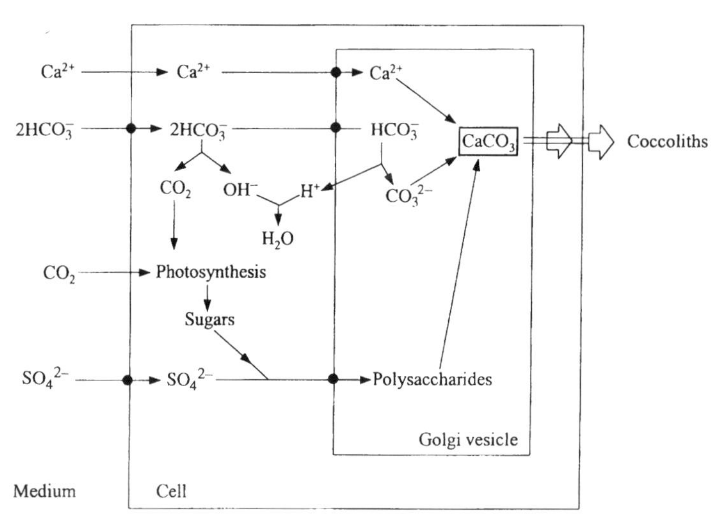

Fig. 5.27 Ion fluxes involved with the intravesicular deposition of calcite in the coccolithophore *Emiliania huxleyi*.

from the edges of the Golgi complex and regroup around the edges of the larger organic scale-bearing vesicles to produce floppy ear-like extensions loaded with calcium and polysaccharide (labelled as CL in Fig. 5.26). Calcium ions are then transported into the chamber of the main vesicle by dissolution of the coccolithosomes followed by active transport through a transmembrane Ca^{2+}-activated ATPase.

In the case of bicarbonate, as a result of the slightly lower intracellular pH and the presence of *carbonic anhydrase*, HCO_3^- ions entering the cell undergo partial dissociation to CO_2 (Fig. 5.27). They are then photosynthetically fixed along with CO_2 molecules that originate from passive diffusion across the cell membrane. As this occurs in the virtual absence of calcium, there is no associated deposition of calcium carbonate unlike in the intercellular spaces of *Halimeda*. However, some HCO_3^- ions manage to escape the pull of photosynthesis and reach the flattened coccolith vesicle where they are transported through the lipid bilayer. As shown in Fig. 5.27, this probably involves a HCO_3^-/H^+ symport that returns the protons dissociated from the transferred HCO_3^- ions back to the cytoplasmic fluid where they neutralize the OH^- ions associated with CO_2 fixation.

Compared with the situation in *Halimeda*, the connection between photosynthesis and calcification is less clear in the coccolithophores. It is clear that photosynthesis influences the rate of calcification in the coccolith vesicle by modulation of the flow of bicarbonate anions, as well as providing metabolic energy for polysaccharide synthesis, base-plate deposition and vesicle assembly. However, other factors, such as the flux of Ca^{2+} ions and rate of delivery of the coccolithosomes, are also very important. Indeed, unlike *Halimeda*, which doesn't do it when the light goes out, naked coccolithophores are capable of performing in the dark! This is particularly prevalent in the exponential growth phase during which the coccoliths are first produced. At this stage it appears that the rate of coccolith formation is about the same in the light and dark. Presumably, as long as there is enough available energy to maintain the Ca^{2+} and HCO_3^- transport systems and matrix synthesis, and that the build up of H^+ ions from HCO_3^- dissociation in the coccolith vesicle is not excessive, then calcification continues unabated.

In coccolithophores, elaborate calcite scales are produced within intracellular vesicles by a complex process of ion fluxes, some of which involve photosynthetic pathways but the interdependence appears not to be essential. Vesicle–vesicle interactions are important in controlling the deposition of an organic base-plate prior to calcification, and Ca^{2+} ions are supplied via polysaccharide-rich calcium-containing nanoparticles called coccolithosomes.

5.5 Summary

In this chapter we have highlighted how the delineation of biological environments provides sites of controlled chemistry that are spatially defined. The sealing off of fluid spaces in boundary-organized biomineralization provides a mechanism for regulating the size, shape and organization of biominerals, as well as controlling the ionic composition and supersaturation required for nucleation and growth. The minerals are usually encased within an insoluble organic shell so they become protected from chemical and biochemical dissolution.

Phospholipid and polypeptide vesicles, cellular assemblies and macromolecular frameworks can be assembled into enclosed semipermeable structures that provide diffusion-limited spaces for biomineralization processes. Several examples, including magnetite crystals in magnetotactic bacteria (phospholipid vesicles), iron oxide nanoparticles in ferritin (polypeptide vesicles), calcium phosphate in certain types of bone (cellular assemblies), and calcite in avian eggshells, goethite in limpet teeth, and aragonite in the nacreous layers of seashells (macromolecular frameworks), were discussed. A common feature is that vesicles and cell membranes are associated with the transport and accumulation of ions such as Ca^{2+}, Fe^{II}, Fe^{III}, H^+, OH^-, HCO_3^- and HPO_4^{2-}, and these processes establish levels of supersaturation compatible with controlled mineralization.

The types of ion-transport systems available for exploitation in biomineralization were illustrated by reference to algal calcification in which ion fluxes directly or indirectly result in localized supersaturation levels within intracellular vesicles (coccolithophores), intercellular spaces (*Halimeda*) or close to the external surface of the cell wall (*Chara*). These examples are complicated by the photosynthetic removal of CO_2, which can interfere with the bicarbonate–carbonate balance in the boundary-organized fluids. The formation of coccolith scales is highly complex because it not only involves ion fluxes across several boundaries but several different types of vesicles.

In conclusion, our perspective in this chapter has been focused on ways of physically partitioning biological space using organic walls with selective permeability. We have considered the effect of this strategy on controlling the solution chemistry (supersaturation) but have not addressed questions concerning the mechanisms of mineral nucleation and growth. As we discussed in Chapter 4, Section 4.4, nucleation is strongly influenced by the presence of an insoluble substrate because this tends to lower the activation energy barrier preventing the formation of stable aggregates in the supersaturated solution. There seems, therefore, every reason to expect that spatial boundaries formed from phospholipid bilayers or macromolecular frameworks will also play an important role in determining the nucleation of biominerals. This novel aspect of structural control is discussed in the next chapter, which describes the general principles of organic matrix-mediated biomineralization.

Further reading

Borowitzka, M. A. (1989). Carbonate calcification in algae—initiation and control. In *Biomineralization: chemical and biochemical perspectives* (ed. Mann, S., Webb, J. and Williams, R. J. P.), pp. 63–94. VCH Verlagsgesellschaft, Weinheim.

de Vrind-de Jong, E. W., Borman, A. H., Thierry, R., Westbroek, P., Gruter, M. and Kamerling, J. P. (1986). Calcification in the coccolithophorids *Emiliania huxleyi* and *Pleurochrysis carterae*. I. Ultrastructural aspects. In *Biomineralization in lower plants and animals* (ed. Leadbeater, B. S. C. and Riding, R.), pp. 189–204. Systematics Association Vol. 30. Oxford University Press, Oxford.

Ford, G. C., Harrison, P. M., Rice, D. W., Smith, J. M. A., Treffry, A., White, J. L. *et al.* (1984). Ferritin: design and formation of an iron-storage molecule. *Philos. Trans. R. Soc. London B*, **304**, 551–565.

Gorby, Y. A., Beveridge, T. J. and Blakemore, R. P. (1988). Characterization of the bacterial magnetosome membrane. *J. Bacteriol.*, **170**, 834–841.

Mann, S. and Frankel, R. B. (1989). Magnetite biomineralization in unicellular microorganisms. In *Biomineralization: chemical and biochemical perspectives* (ed. Mann, S., Webb, J. and Williams, R. J. P.), pp. 389–426. VCH Verlagsgesellschaft, Weinheim.

Silyn-Roberts, H. and Sharp, R. M. (1986). Crystal growth and the role of the organic network in eggshell biomineralization. *Proc. R. Soc. London B*, **227**, 303–324.

Stryer, L. (1988). *Biochemistry*. W. H. Freeman, New York.

Webb, J., Macey, D. J. and Mann, S. (1989). Biomineralization of iron in molluscan teeth. In *Biomineralization: chemical and biochemical perspectives* (ed. Mann, S., Webb, J. and Williams, R. J. P.), pp. 345–438. VCH Verlagsgesellschaft, Weinheim.

Weiner, S. (1986). Organization of extracellularly mineralized tissues: a comparative study of biological crystal growth. *Crit. Rev. Biochem.*, **20**, 365–408.

Westbroek, P., Van der Wal, P., van Emburg, P. R., de Vrind-de Jong, E. W. and de Bruijn, W. C. (1986). Calcification in the coccolithophorids *Emiliania huxleyi* and *Pleurochrysis carterae*. II. Biochemical aspects. In *Biomineralization in lower plants and animals* (ed. Leadbeater, B. S. C. and Riding, R.), pp. 205–217. Systematics Association Vol. 30. Oxford University Press, Oxford.

6 Organic matrix-mediated biomineralization

The wide-scale involvement of insoluble organic structures in controlled biomineralization is referred to as *organic matrix-mediated biomineralization*. The organic matrix is specifically synthesized under genetic control before and sometimes during the biomineralization process. In most cases, the matrix is a polymeric framework that consists of a complex assemblage of *macromolecules*, such as proteins and polysaccharides.

There are several functions associated with the organic matrix, including:

- *mechanical design*—the modification of physical properties such as strength and toughness to meet the demands of biological activity.
- *mineral passivation*—the surface stabilization of minerals from dissolution or phase transformation.
- *mineral nucleation*—the control of the location and organization of nucleation sites, and the structure and crystallographic orientation of the inorganic phase.
- *spatial delineation and organization*—partitioning of microenvironments with semipermeable frameworks for controlled growth (see Chapter 5, Section 5.1.4).

In this chapter, we focus principally on the types of structural organic frameworks that are found in biomineralization and their use as substrates for the controlled nucleation of inorganic minerals. The notion of organic matrix-mediated biomineralization is also strongly associated with the unique mechanical properties of biominerals, so we begin the chapter with a short section on the mechanical design of macromolecular frameworks. A more detailed discussion of this topic can be found elsewhere (see further reading).

The organic matrix is a preformed insoluble macromolecular framework that is a key mediator of controlled biomineralization. The matrix subdivides the mineralization spaces, acts as a structural framework for mechanical support, and is interfacially active in nucleation.

6.1 Organic matrices as mechanical frameworks

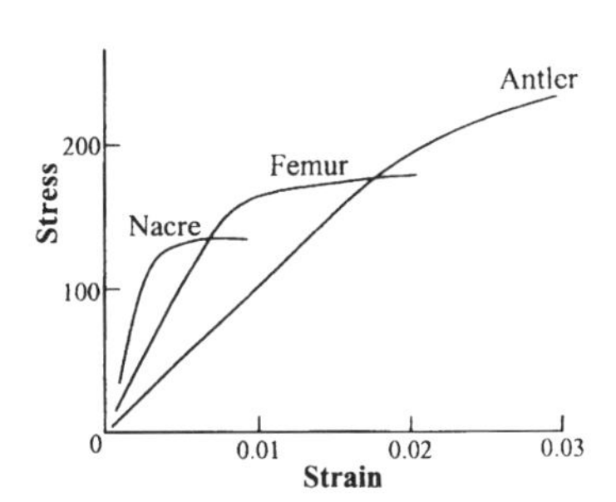

Fig. 6.1 Stress–strain curves for biomineral composites.

Scientists have been interested for many years in the mechanical design of biominerals, such as bone and shell, which are complex composites of inorganic minerals and organic macromolecules that together exhibit unusual toughness, strength and hardness. We have mentioned some general biomechanical features of shells and bone in Chapter 2 (Sections 2.1.1 and 2.3.1, respectively), and Fig. 6.1 shows how these materials can differ in their *stress–strain* properties depending on their skeletal type. Stress is measured in units of pressure (force/area) and can reach a level of around 150 MPa for

Table 6.1 Mechanical properties of bone

	Tension Strength (MPa)	Modulus (GPa)	Compression Strength (MPa)	Modulus (GPa)
Normal bone	130	17	150	9
Bone without organic matrix	6	17	40	7

nacre, and 200 MPa for a typical thighbone (femur) before breakage occurs. Strain, on the other hand, is the percentage extension in length associated with the applied stress, and can attain relatively high values in a material such as antler, which continues to extend but not break even at stresses of 300 MPa. This property is extremely useful if male deer are to maintain a sense of dignity during the head-butting season.

Removing the organic matrix from these biominerals has a significant effect on the mechanical properties. In bone, for example, the strength in tension or compression drops by 95 and 73 per cent, respectively (Table 6.1). However, the stiffness, which is the stress/strain ratio or *Young's modulus* (E), does not change significantly because it is determined principally by the mineral phase.

In general, the incorporation of mineral crystallites within a structural framework of hydrophobic macromolecules must even in the simplest support system resist the forces of tension, compression and bending present in the biological environment. The best way of organizing spatial structures that are braced to withstand external forces is well known (we hope) among architects and civil engineers. Usually it is assumed that the design of the framework involves the property of *least weight* in which the minimal amount of material is used to perform the appropriate function. If this principle is applied to the simplified case of two dimensions, then the theoretical results predict that the optimum structures are in the form of a series of nets as shown in Fig. 6.2. Although the rectangular net (Fig. 6.2A) wins out, the results show that the equiangular spiral (Fig. 6.2B) and circular fan (Fig. 6.2C) also come very close to the absolute minimum weight. Indeed, except for restrictions at the origin of the spiral and at the closure of the fan where strains may be incompatible, these frameworks are remarkably effective. A further conclusion of

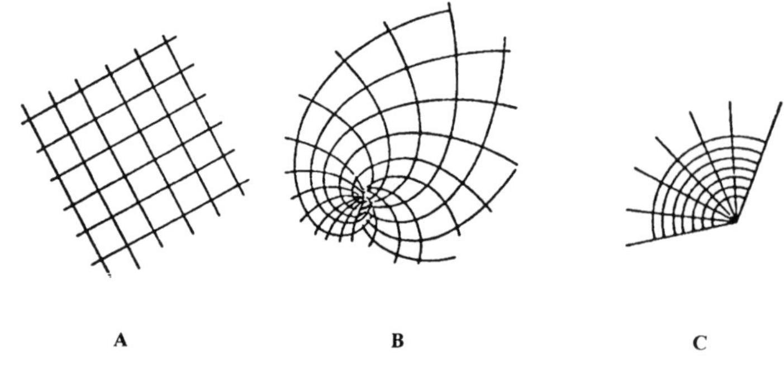

Fig. 6.2 Spatial nets.

the theory is that the optimum framework of minimum weight is also the stiffest of all the possible structures whose members sustain tension and compression stresses in that region of space.

Based on these arguments, it seems most likely that there are many possible solutions to the mechanical design of the structures of the organic matrix which use the least weight approximation and become fine-tuned by the particular stresses placed on the frameworks during and after their synthesis. Whereas the rigidity of the matrix requires cross-linking of strands of connecting fibres or stacks of sheets, the stiffness may not be optimized because the mineralized components provide the required strength and hardness. In fact, a matrix that was intentionally less stiff could have certain advantages—for example as a shock absorber—as it is designed to accommodate rather than resist deformations. This appears to be the case for the iron oxide teeth of limpets that we described in Chapter 2, Section 2.6.3 (see also further reading). The chisel-like cutting edge is produced by the constant rasping of the tooth across the surface of the rock and is supported by an underlying organic matrix of fibres that at the macroscopic scale run in alternate directions, as shown in Fig. 6.3. The fibres are aligned so as to maximize the cutting potential of the sabre-like tooth. In particular, thicker organic cables lie behind the cutting edge where they can absorb the maximum amount of energy associated with the shock wave accompanying each rasping action. See Chapter 5 (Section 5.1.4 and Fig. 5.16) for details of the microscopic structure of the matrix.

Organic frameworks play an important role in the mechanical design of biomineralized tissues such as bones, shells and teeth. Many of the general functions of these biominerals—movement, protection, cutting and grinding—are dependent on mechanical properties, such as strength and toughness, which are specifically associated with inorganic–organic composites.

6.2 Macromolecules and the organic matrix—a general model

Organic matrices are produced as insoluble frameworks in aqueous environments so it follows that their constituent macromolecules must be predomi-

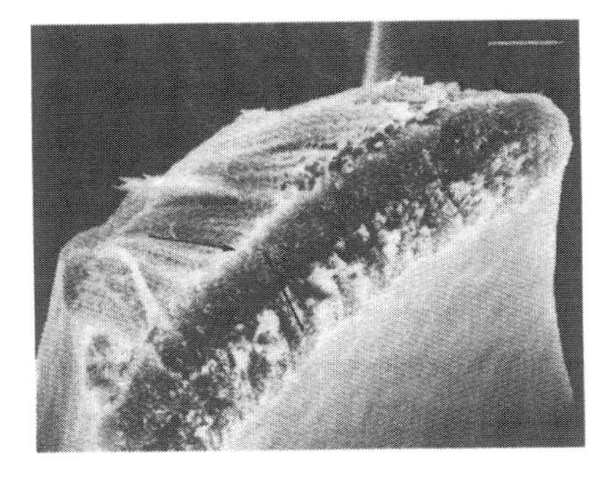

Fig. 6.3 Tip of a limpet tooth showing internal organic matrix. Scale bar, 10 μm.

nantly hydrophobic; otherwise, the framework would readily redissolve with time. Clearly, this hydrophobic character is necessary if the matrix is to be employed as a semiporous net for delineating space or as a mechanical frame. But it seems at odds with the notion that such frameworks are of central importance in controlling the nucleation of biominerals, because this involves the organization of ions and hydrated species, which under normal circumstances would not interact strongly with a hydrophobic surface. The competing requirements associated with this hydrophilic–hydrophobic balance can be met in several ways, for example by

- *amphiphilic* macromolecules
- *cross-linking* of a relatively soluble matrix after initial deposition
- *multicomponent* systems that contain macromolecules with different hydrophobic/hydrophilic properties.

In recent years, a general model that accounts for both the *structural* and *functional* aspects of the organic matrix has been proposed. In its simplest form the matrix consists of a structural framework of predominantly hydrophobic macromolecules with associated cross-links, onto which are anchored hydrophilic macromolecules that present an active nucleating surface to the external solutions (Fig. 6.4).

The two-component model has arisen primarily from empirical evidence based on the extraction of proteins from mineralized extracellular tissues such bone and shells (see further reading). In these studies, two classes of macromolecules—*framework macromolecules* and *acidic macromolecules*—have been identified in terms of their low and high solubilities in water, respectively, after the biomineral has been dissolved by chemical methods. This often entails the use of a mixture of hydrochloric acid and a chelating agent such as ethylenediaminetetraacetic acid (EDTA) for calcium carbonate and calcium phosphate, and concentrated nitric and sulfuric acids followed by aqueous hydrofluoric acid for silica. The framework macromolecules are associated with the insoluble fraction and consist of hydrophobic proteins and polysaccharides, and are thought to represent the structural and biomechanical components of the organic matrix. In contrast, the numerous water-soluble acidic macromolecules released during the demineralization process correspond to functional components involved in controlling nucleation. These macromolecules might also play a role in controlling mineral growth and general cellular activities associated with biomineralization, such as ion transport, enzymatic regulation and hormonal signalling.

Some researchers have attempted to isolate additional fractions from the heterogeneous mixture of insoluble framework proteins. For example, further

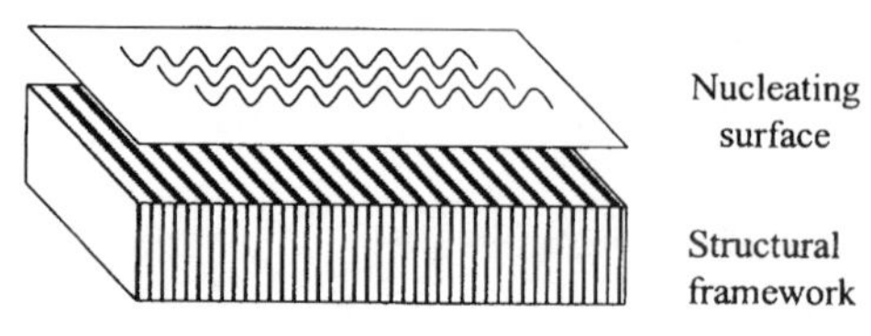

Fig. 6.4 Two-component model of the organic matrix.

treatment with weak alkaline solutions can sometimes result in the release of macromolecules into solution. As these components are strongly attached to the framework under normal conditions, they are likely to represent proteins that have an *interfacial* role, for example in linking together the framework and acidic macromolecules.

Some of the main types of macromolecules known to be associated with organic matrix-mediated biomineralization are shown in Table 6.2. Framework macromolecules such as collagen, chitin and cellulose have been extensively studied because of their general importance in biology. Collagen in particular is of key importance in the biomineralization of bone and we discuss the remarkable structure of this protein in Section 6.3.1 Our knowledge of the acidic macromolecules, in contrast, is scarce because they are often difficult to isolate as a pure single protein, and even when this is possible, the amino acid sequence is not easy to obtain. Often the best we can hope for is an amino acid composition that provides a broad overview of the type of chemical groups available for interaction with mineral ions present in a supersaturated solution. In fact, it is on this basis that the protein macromolecules are defined as 'acidic' since they often contain large numbers of aspartic acid and glutamic acid residues, and serine and threonine amino acids that are modified with covalently bound phosphate groups. In many cases, the acidic macromolecules are *glycoproteins*, which are proteins with covalently linked polysaccharide side chains that often containing sulfate and carboxylic acid

Table 6.2 Main types of macromolecules associated with organic matrix-mediated biomineralization

System	Framework macromolecules	Acidic macromolecules
Bone and dentine	Collagen	Glycoproteins Osteopontin Osteonectin Proteoglycans Chondroitin sulfate Keratin sulfate Gla-containing proteins Osteocalcin
Tooth enamel	Amelogenin	Glycoproteins Enamelins
Mollusc shells (nacre)	β-Chitin Silk-like proteins (MSI 60) N16/N14? Lustrin A	Glycoproteins Nacrein N66
Crab cuticle	α-Chitin	Glycoproteins
Diatom shells	Frustulins	Glycoproteins HEP200 Silaffins
Silica Sponges	Silicatein	No data
Plant silica	Cellulose	Proteins Carbohydrates (xylose, glucose)

residues. Details of these and other functional groups associated with organic matrices in biomineralization are described in Sections 6.3 to 6.6.

As a general model, the organic matrix consists of a structural framework of hydrophobic macromolecules in association with acidic macromolecules that act as a nucleation surface for biomineralization. The two components, which often contain a complex assemblage of macromolecules, can be extracted in the laboratory by dissolution of the mineral phase.

6.3 Matrix macromolecules in bone

The matrix macromolecules of bone comprise an insoluble framework of cross-linked *collagen* fibrils in association with *non-collagenous* proteins that are soluble in water. Collagen accounts for approximately 90 per cent by weight of the total protein content, whereas 200 different non-collagenous proteins are found in the remaining 10 per cent! Many of these originate from the circulating blood and become adventitiously trapped in the bone matrix so they are not directly connected with biomineralization. Those deemed to be relevant are often given exotic names, several of which are mentioned in Section 6.3.2. Sometimes the same protein has two or more names, which causes great confusion and lots of arguments at conferences.

Fortunately, although 11 types of collagen are known in biology, only the major form, *type I*, is produced by bone-forming cells (*osteoblasts*). Type I collagen fibrils have a specific structure that arises from a series of steps that occur initially inside the osteoblast and then proceed outside the cell in the surrounding extracellular space. Inside the osteoblast, the steps include:

synthesis of helical polypeptide chains
↓
enzymatic modification of amino acids (proline and lysine hydroxylation)
↓
self-assembly of triple-stranded helix filaments
↓
secretion into the extracellular space

Once in the extracellular environment, the nascent collagen undergoes several further stages of processing:

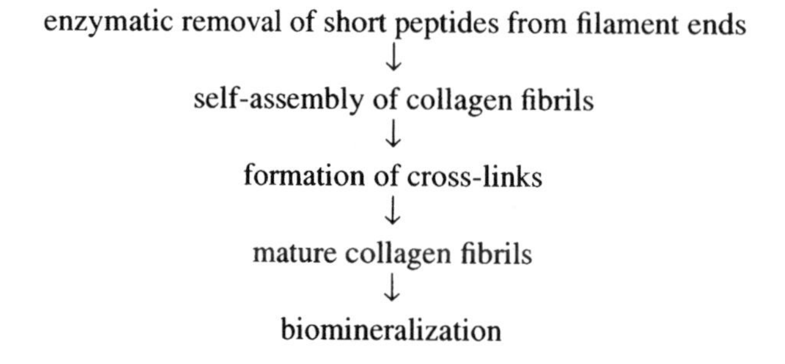

Overall, the process is a remarkable example of hierarchical self-assembly in which each stage involves building blocks of increasing size such that the structure is organized across a range of length scales.

The organic matrix of bone consists predominantly of collagen in association with a wide range of non-collagenous proteins. The assembly of collagen fibrils involves a series of steps beginning inside the osteoblast and ending in the extracellular space.

6.3.1 Collagen

At the molecular level, collagen consists of about 1000 amino acids arranged along a helical polypeptide backbone. As shown in Table 6.3, over 30 per cent of the residues are *glycine* (Gly) (Fig. 6.5), which is highly unusual as most proteins, for example haemoglobin, have only about 5 per cent of this amino acid. Collagen also contains significant amounts of the amino acid *proline* (Pro), often in the form of *4-hydroxyproline* (Hyp) (Fig. 6.5). Another unusual amino acid, *5-hydroxylysine* (Hyl), is also found in collagen (Fig. 6.5). The modified amino acids are synthesized after assembly of the peptide chain by hydroxylation of the proline and lysine residues, respectively, by special enzymes in association with a reducing agent such as ascorbic acid.

One of the most distinctive features about collagen is that the glycine residues are regularly spaced along the helical backbone to give 338 contiguous repeats of the triplet sequence,

[Gly–X–Y]

Table 6.3 Amino acid composition (%) of rat tail tendon

Alanine	9.9	Threonine	1.9
Glycine	35.1	Methionine	0.6
Valine	2.3	Arginine	4.7
Leucine	2.2	Histidine	0.3
Isoleucine	1.3	Lysine	3.6
Proline	12.3	Aspartic acid	4.7
Hydroxyproline	9.0	Glutamic acid	7.4
Tyrosine	0.5	Serine	2.8
Phenylalanine	1.4		

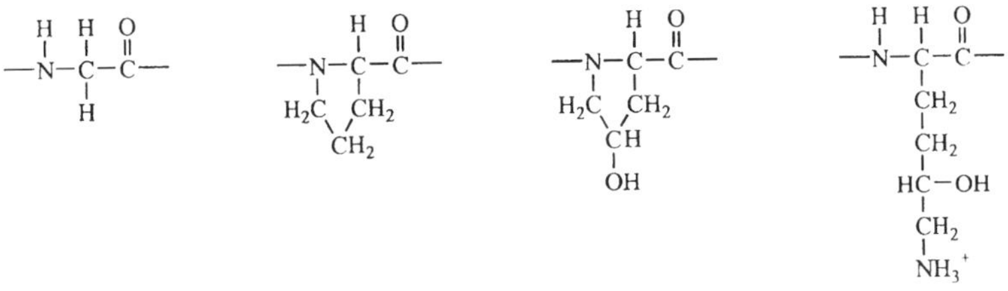

Gly (G) **Pro (P)** **Hyp** **Hyl**

Fig. 6.5 Characteristic amino acid side chains found in collagen.

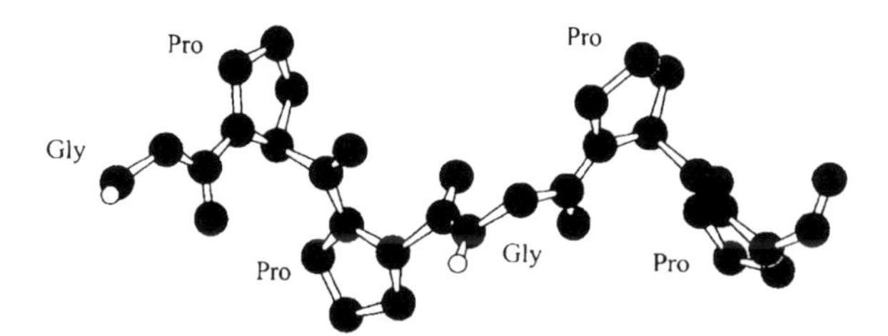

Fig. 6.6 Helical twisting of collagen polypeptide chain.

in which every third amino acid is glycine and X and Y vary along the chain but are typically proline (Pro) and 4-hydroxyproline (Hyp), respectively. The repetition of the sequence [Gly-Pro-Hyp] has two major consequences. First, the steric constraints associated with the pyrrolidone rings of the proline residues cause a regular twisting of the polypeptide chain into an open helical form (Fig. 6.6). Second, the helices can readily pack tightly together to form a supramolecular aggregate because the single hydrogen atom of the glycine side chain does not take up much space. It turns out that efficient packing is achieved when three molecules wrap round each other to produce a superhelical cable, referred to as *tropocollagen*.

Figure 6.7 shows the coiled-coil structure of the tropocollagen filament in terms of the α-carbon backbone of the polypeptide chains. Each individual chain coils around a slightly different axis (Fig. 6.7A) so that the triple-stranded helix consists of three *non-coaxial* helical polypeptides wound together along a common direction (Fig. 6.7B). Viewed end-on, the structure clearly shows how the glycine residues, shown as filled circles in Fig. 6.7C, are internalized within the fibril due to their small size and compact space-filling. In contrast, the bulky pyrrolidone rings of the proline and hydroxypro-

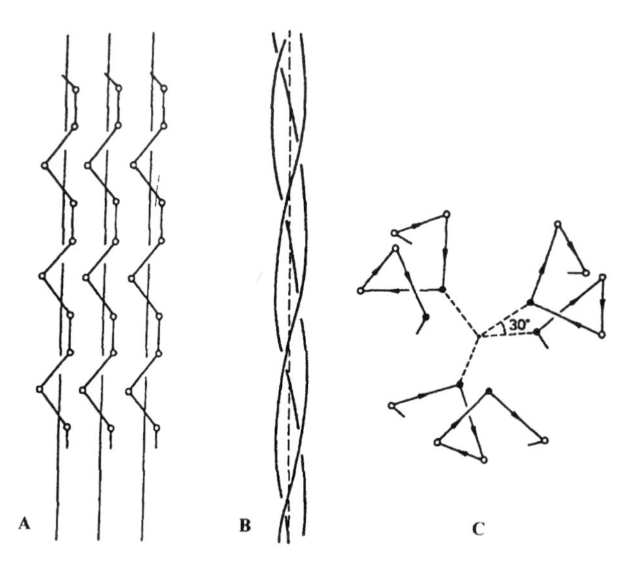

Fig. 6.7 Triple helical structure of tropocollagen filament: (A) helical polypeptide chains; (B) super-helix; (C) end-on view along axis of superhelix.

line residues that are located either side of the glycine side chains are positioned on the outside of the filament. The number of residues per turn of the superhelix is 3.3—hence the requirement for every residue in three to be glycine—and the distance per turn is 0.3 nm. Overall, the triple-stranded helix is 280 nm in length and 1.5 nm in width, and has a molecular mass of around 285 000.

In bone, two of the polypeptide chains in the tropocollagen filament are identical whereas the third is different, although all three contain 338 tandem repeats. This arrangement is further stabilized by several types of *interchain* interactions, such as

- *steric locking* of the proline and hydroxyproline residues.
- *hydrogen bonding* between peptide -NH donors of the glycine residues of one chain with peptide -CO hydrogen acceptors on the other chains, which leads to hydrogen bonds that lie perpendicular to the long axis of the tropocollagen filament.
- *hydrogen bonding* between hydroxyl groups of the hydroxyproline residues and bridging water molecules present in the superstructure.
- *covalent cross-links* derived from the aldol condensation of lysine side chains located in the non-helical region close to the amino terminus (Fig. 6.8).

Together, these interactions produce triple-stranded helical filaments of uniform size, composition and structure that readily form fibrous crystals in aqueous salt solutions.

Crystals of type I collagen consist of a specific fibril structure, called the *revised quarter-stagger* model, in which the tropocollagen filaments are lined up head-to-tail in rows that are staggered by 64 nm along their long axis (Fig. 6.9). Each filament in Fig. 6.9 is divided into five zones, the first four are equal in length (64 nm) but the fifth is much shorter (25 nm). The filaments are transposed by 64 nm along their length because this arrangement maximizes the number of strong *interfilament cross-links*, which

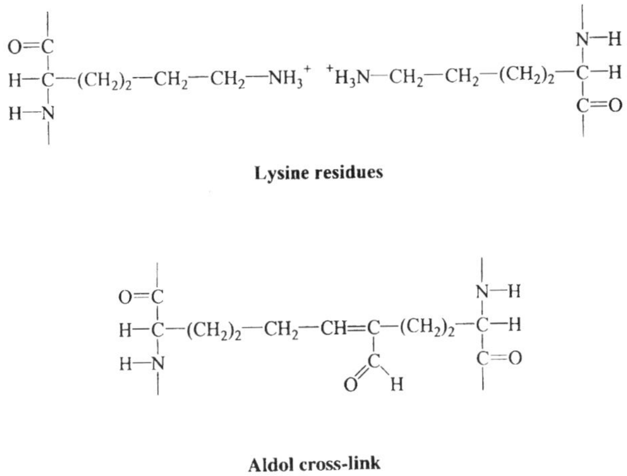

Fig. 6.8 Aldol cross-link from two lysine side chains.

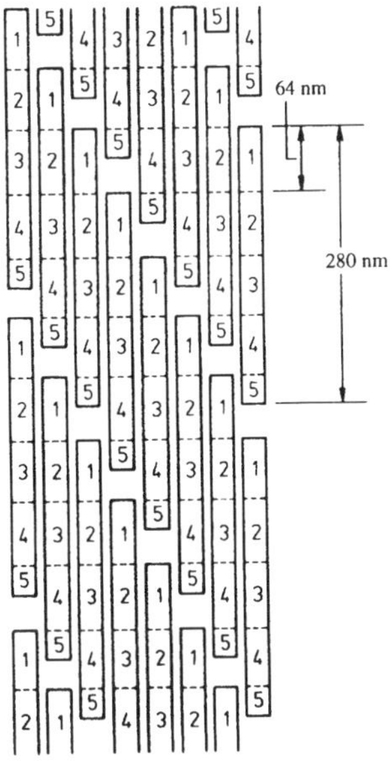

Fig. 6.9 Revised quarter-stagger model of collagen fibrils.

increase the stability of the crystal lattice. The nature of the covalent cross-links is complex but generally derives from the coupling of hydroxylysine and lysine residues to produce *hydroxypyridinium* bridges. For example, Fig. 6.10 shows one such bridge formed from the coupling of two hydroxylysine residues and one lysine residue. The cross-links are precisely located between the triple-stranded helices and a particularly important connection occurs between the amino-terminus of one filament and those close to the carboxy-terminus of an adjacent superhelix (Fig. 6.11). It is this linkage that gives rise to the mismatch in the sizes of zones 1 (N-terminus) and 5 (C-terminus) shown in Fig. 6.9. Moreover, as shown in Fig. 6.9, the arrangement results in the formation of a regular array of small gaps between filaments aligned within the same row. The spaces are 40 nm in length and 5 nm in width, and usually referred to as the *hole zones*. Detailed studies of the three-dimensional structure of the crystalline collagen fibrils—see further reading for details—suggest that adjacent hole zones overlap to produce grooves that are organized in parallel rows along the long axis of the structure.

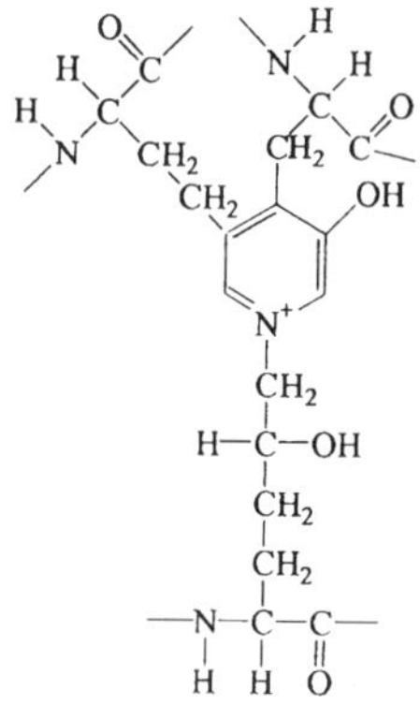

Fig. 6.10 Hydroxypyridinium bridge in collagen.

The importance of the hole zones and associated grooves in bone mineralization is well recognized. These sites are considered to be the loci for the specific nucleation and growth of crystals of hydroxyapatite that are organized in bands across the collagen fibrils. The crystals are initially confined to the hole zones so they grow into plate-shaped particles, 45 nm in length, 20 nm in width and only 3 nm in thickness. The crystals are therefore very thin with large flat faces that lie perpendicular to the $[1\bar{1}0]$ axis. As the unit cell is 0.937 nm in length along this direction, the crystals are only between 2 and 5 unit-cells thick! Moreover, individual crystals are oriented such that the crystallographic *c* and *a* axes are preferentially aligned along the collagen fibril axis and groove direction, respectively (Fig. 6.12). On a longer length scale, the crystals are aligned in parallel arrays across individual fibrils and sometimes this arrangement is even coherent across several adjacent fibrils. Such long-range ordering of the collagen matrix and hydroxyapatite crystallites could be responsible for the unusual fracture properties of bone.

The amino acid composition of type I collagen consists of over 30 per cent glycine (Gly) and unusual residues such as proline (Pro), 4-hydroxyproline (Hyp) and 5-hydroxylysine. The polypeptide chain is helical and consists of a tandem repeat of $[Gly\text{–}X\text{–}Y]_{338}$, *often as [Gly-Pro-Hyp], along the chain. Three non-coaxial helical polypeptides self-assemble to produce the triple-stranded helix of tropocollagen which is further stabilized by interchain cross-linking. Collagen fibrils are constructed from a staggered periodic arrangement of cross-linked tropocollagen filaments that contains regu-*

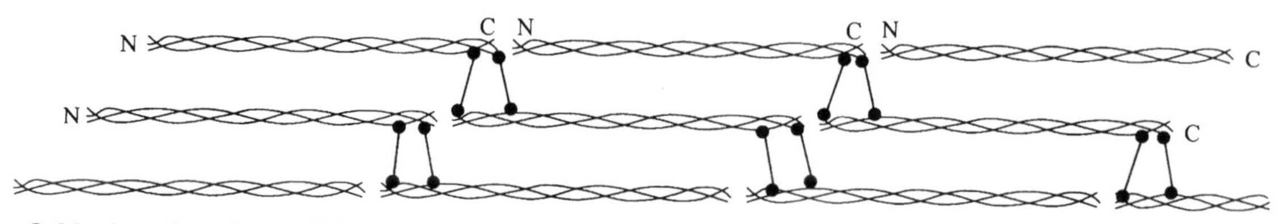

Fig. 6.11 Location of cross-links between the carboxy (C) and amino (N) terminal regions of adjacent tropocollagen filaments in collagen fibrils.

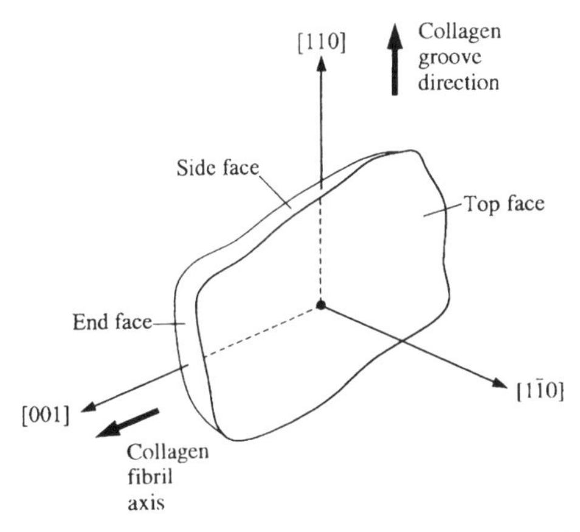

Fig. 6.12 Alignment of plate-shaped hydroxyapatite crystal in the hole zone of the collagen matrix. The [001] and [110] directions correspond to the crystallographic *c* and *a* axes, respectively.

larly spaced hole zones and grooves in which plate-like crystals of hydroxyapatite nucleate and grow.

6.3.2 Non-collagenous proteins in bone

The water-soluble fraction of the organic matrix of bone contains several different types of macromolecules specific to bone (Table 6.4, and see further reading). Many of these proteins are also found in dentine. In general, the macromolecules include:

- *acidic glycoproteins*—proteins enriched in aspartate and glutamate and covalently linked polysaccharide side chains.
- *Gla proteins*—proteins enriched with the modified amino acid *γ-carboxyglutamic acid* (Gla).
- *proteoglycans*—polysaccharide-based structures containing a central core protein that self-assemble into micrometre-scale aggregates.

Table 6.4 Main types of non-collagenous proteins associated with bone and dentine biomineralization

Macromolecules	Molecular mass ($\times 10^3$)	Composition
Acidic glycoproteins		
Osteonectin	44 (bovine)	Aspartate- and glutamate-rich
Sialoprotein II	200	Glutamate-rich
Phosphoprotein	40	Aspartate/glutamate-rich/phosphorylated
Phosphophoryns (dentine)	100 (human)	Aspartate and phosphoserine-rich
Proteoglycans		
Bone proteoglycans	350	Chondroitin sulfate chains
Cartilage proteoglycans	1000	Chondroitin/keratin sulfate chains
Gla proteins		
Osteocalcin	6	γ-Carboxyglutamic acid residues (× 3)
Matrix Gla protein	15	γ-Carboxyglutamic acid residues (× 5)

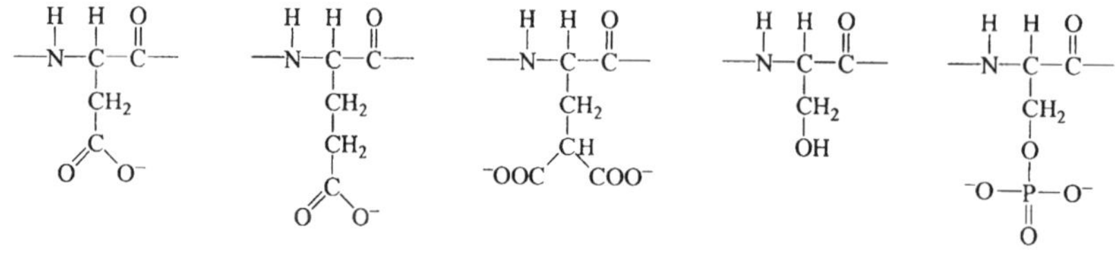

Fig. 6.13 Amino acid side chains associated with acidic macromolecules in biomineralization.

Acidic glycoproteins, such as *osteonectin*, which accounts for about 2.5 per cent by weight of the matrix, often consist of relatively large amounts of deprotonated aspartic acid (Asp) and glutamic acid (Glu) residues (Fig. 6.13) that occur in repeated sequences along the polypeptide chain. For example, in *sialoprotein II* there are runs of glutamates in different parts of the macromolecule, such as

- residues 77 to 84 $[Glu]_3Gly[Glu]_4$
- residues 155 to 164 $[Asp][Glu]_9$.

Similarly, in a protein referred to as *osteopontin*, a run of nine aspartates ($[Asp]_9$) occurs between residues 70 and 78.

Other glycoproteins are best described as *phosphoproteins* because along with the aspartate and glutamate domains they also contain significant numbers of the hydroxy amino acid serine (Ser), modified with covalently linked phosphate groups to give a phosphorylated derivative (PSer) (Fig. 6.13). A supercharged example of this type of macromolecule is found associated specifically with the collagen matrix of dentine. These proteins are referred to as *phosphophoryns* and consist of repeat sequences such as

- $[Asp]_6[Tyr][Ser\text{-}Asp]_2[Ser][Ser\text{-}Asp]_2[Asp]$
- $[PSer]_8$

The precise role that these acidic macromolecules play in bone mineralization is not yet determined. Proteins such as osteonectin, as well as the phosphoproteins, are known to bind to collagen, possibly at particular sites in the hole zones. Clearly, the high numbers of anionic groups will interact strongly with Ca^{2+} ions, so we have a system that fits the two-component structure–function general model described above in Section 6.2. The bone *Gla protein*, *osteocalcin*, also fits into this scheme because each macromolecule contains doubly charged dicarboxylate side chains associated with three separate γ-carboxyglutamate (γ-Glu) residues (Fig. 6.13), each of which have strong calcium-binding properties.

Proteoglycans only account for around 1 per cent by weight of the bone matrix, but are more common in *mineralized cartilage*, which also contains collagen and large amounts of water (60 to 80 per cent). Cartilage is an extracellular matrix that has high tensile strength and resilience. In the early stages of bone growth, it acts as a shape or mould around which the major bones, except the skull, form—see Chapter 8, Section 8.1, for more details. And in mature bones, cartilage covers the mineral surfaces producing smooth and cushioned movements that enable squash players and the like to recklessly charge around in search of a small rubber ball.

A proteoglycan aggregate consists of an organized complex of proteins and long linear polysaccharide chains. The latter are referred to as *glycosaminoglycans* because they are constructed from disaccharide repeating units containing a derivative of an amino sugar, either glucosamine or galactosamine. At least one of the sugars in the repeating disaccharide unit contains a negatively charged group such as carboxylate or sulfate. The aggregate has a huge molecular mass—about 2 million—so the structure not surprisingly is highly complex. At the micrometre scale, the aggregate looks a bit like the end of a tiny bottlebrush (Fig. 6.14). The central filament is made of *hyaluronic acid*, which is a polymer of glucuronic acid and *N*-acetylglucosamine (Fig. 6.15). Over 100 proteoglycan monomers are non-covalently coupled at regular intervals of 30 nm to opposite sides of the central hyaluronate backbone by a linker protein. Each proteoglycan monomer consists of a central core protein, 300 nm in length, along which approximately 80 glycosaminoglycan chains are covalently linked through serine and threonine residues.

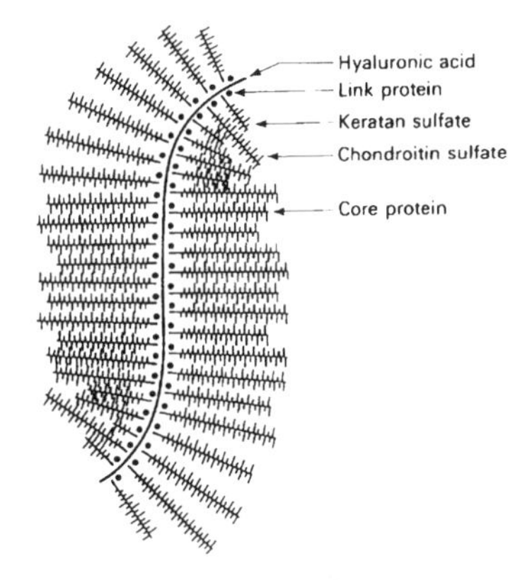

Fig. 6.14 Proteoglycan aggregate.

The glycosaminoglycan side chains come in two types—*chondroitin sulfate* and *keratan sulfate*—with molecular masses around 20 000 (Fig. 6.15). The former is a polymer of glucuronic acid and sulfated *N*-acetylglucosamine and has an anionic group on each saccharide ring (carboxylate and sulfate, respectively), while the latter has a disaccharide repeat unit of galactose and sulfated *N*-acetylgalactosamine with a single negative charge. Clearly, the chains are highly anionic and bind large numbers of Ca^{2+} ions.

Non-collagenous proteins in bone include acidic glycoproteins (osteonectin, sialoprotein II, osteopontin, phosphoproteins), Gla proteins and proteoglycans. The glycoproteins are enriched in anionic residues (aspartate, glutamate, phosphoserine) that are often repeated in certain regions of the polypeptide chain. The Gla protein, osteocalcin, contains three γ-carboxyglutamate residues. Proteoglycans monomers, in contrast, are polysaccharide-based structures consisting of a protein core and 80 or so covalently linked glycosaminoglycan (chondroitin sulfate, keratan sulfate) side chains. The monomers in turn are non-covalently linked to a central hyaluronate filament to produce a highly branched aggregate several micrometres in size.

6.4 Tooth enamel proteins

Unlike dentine, which consists of a mineralized collagenous matrix in the centre of the tooth, mature enamel contains only small amounts (< 5 per cent)

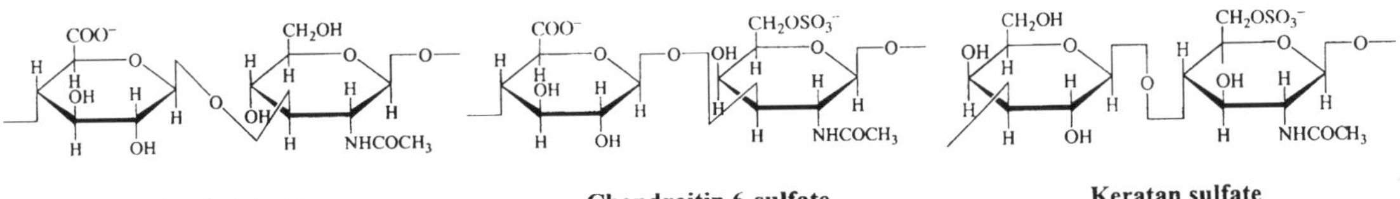

Hyaluronate **Chondroitin 6-sulfate** **Keratan sulfate**

Fig. 6.15 Glycosaminoglycans associated with proteins in proteoglycan aggregates.

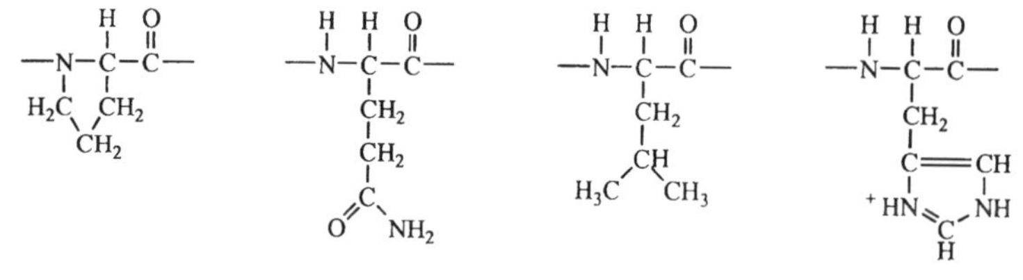

Pro **Gln** **Leu** **His**

Fig. 6.16 Characteristic amino acids found in amelogenins.

of organic macromolecules. The matrix is secreted by specialized cells—*ameloblasts*—and as for bone, the macromolecules can be divided into hydrophobic water-insoluble components and acidic glycoproteins.

The hydrophobic proteins are referred to as *amelogenins*, and although present in relatively large amounts in the early stages of enamel deposition, they are degraded and removed as the hydroxyapatite crystals develop. The proteins typically consist of 180 amino acids, rich in proline (Pro), glutamine (Gln), leucine (Leu) and histidine (His) (Fig. 6.16). Several amino acid sequences have been determined (see further reading) and although these are highly conserved there are no regular repeating domains along the polypeptide chain. Although the individual protein molecules are relatively small—the molecular mass is only about 25 000—approximately 100 monomers self-assemble into uniform-sized 20-nm *nanospheres* that may play an important spatial role in controlling the growth of hydroxyapatite crystals (Fig. 6.17). The nanospheres bind to specific crystallographic planes of the growing crystallites such that these surfaces are blocked from further growth and the crystals develop preferentially along the *c* axis. Two specific domains along the polypeptide chain, between residues 1 and 42, and 157 and 173,

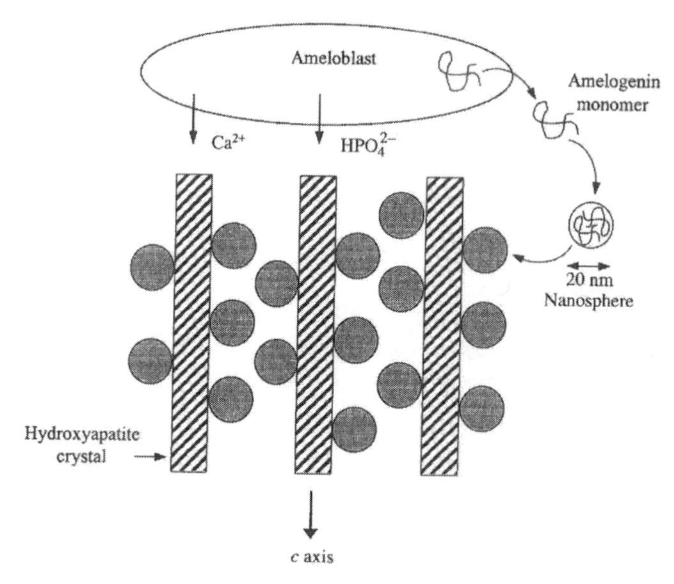

Fig. 6.17 Oriented growth of enamel crystals in a matrix of amelogenin nanospheres.

appear to be particularly important in this process, which results in the growth of long spaghetti-shaped hydroxyapatite filaments.

Enamelins, in contrast, are a family of highly acidic glycoproteins, rich in glutamate and aspartate, as well as glycine residues, with a molecular mass typically between 56 000 and 67 000. The enamelins form a sheath or tubule around the growing hydroxyapatite crystallites. X-ray diffraction studies (see further reading) indicate that the polypeptide chain adopts a specific type of secondary structure, referred to as a *β-pleated sheet*. This conformation is very different from the α-helical arrangement often found in proteins—for example in ferritin (Chapter 5, Section 5.1.2)—and is described in more detail in the next section.

Enamel contains both hydrophobic (amelogenins) and acidic (enamelins) macromolecules in low concentrations. Although there is no extended organic framework, the amelogenins form nanospheres that spatially control the crystal growth process. The polypeptide chain of enamelins adopts a β-pleated sheet secondary structure.

6.5 Matrix macromolecules from shell nacre

In *nacre*, the organic matrix is in the form of a tough organic sheet that is assembled in parallel stacks between ordered arrays of plate-like aragonite crystals—see Chapter 5 (Section 5.1.4, Fig. 5.17) for details. Each sheet is a composite of several components organized as shown schematically in Fig. 6.18. At the centre of each sheet is a hydrophobic framework of insoluble macromolecules with structures and compositions similar to proteins found in silk fibres. They are therefore usually referred to as *silk fibroin-like* proteins. In some types of nacre, fibrils of the straight-chain polysaccharide *β-chitin*

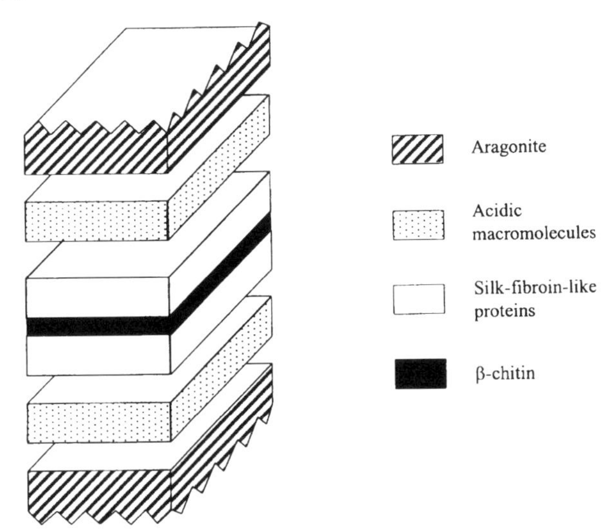

Fig. 6.18 Cross-section through a single matrix sheet and adjacent aragonite crystals in nacre. The crystals are oriented with their *c* axes perpendicular to the organic surface.

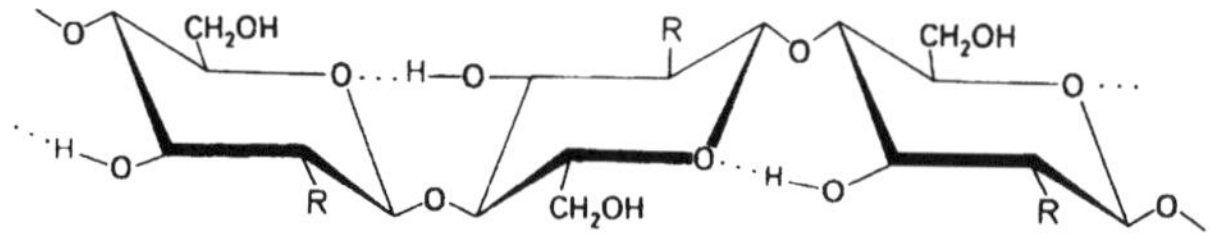

Fig. 6.19 Structure of cellulose (R = -OH) and chitin (R = $-NHCOCH_3$).

are sandwiched between layers of the hydrophobic proteins. β-Chitin, which is closely related to cellulose, is a polymer of *N*-acetylglucosamine in which individual chains are connected through β-1,4-linkages (Fig. 6.19) and adjacent chains run in antiparallel directions. Acidic macromolecules are intimately associated with the surface of the silk-like proteins such that the organic sheets are functionalized with charged and polar residues that can induce the oriented nucleation of aragonite from the surrounding fluid.

The silk fibroin-like proteins have an *antiparallel β-pleated sheet* secondary structure that consists of polypeptide chains containing large amounts of glycine (Gly) and alanine (Ala) (Fig. 6.20). The antiparallel β-pleated sheet is formed when a polypeptide chain loops back on itself such that two adjacent segments of the chain are locked together by strong interchain hydrogen bonds between facing -CO and -NH groups of the peptide backbone. The result is a small corrugated sheet that can be extended laterally by further loops in the polypeptide that reverse the chain direction and bring sections of the backbone back into register with the pleated domain. This is illustrated in the top face (running along the *a* axis) of the structure drawn in Fig. 6.21, which also shows how the antiparallel sheets pack in three dimensions in a silk fibre. The large and small circles shown in Fig. 6.21 represent

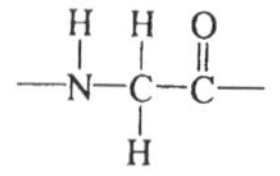

Gly

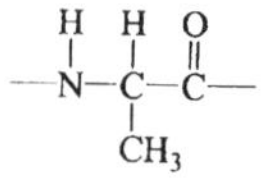

Ala

Fig. 6.20 Characteristic amino acids found in silk-fibroin-like proteins.

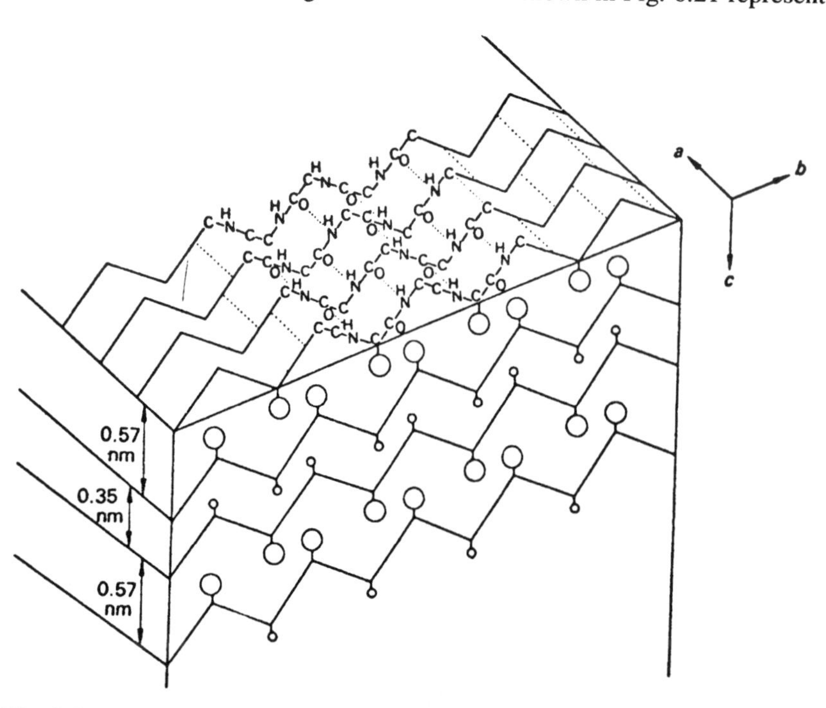

Fig. 6.21 Antiparallel β-pleated sheet structure in silk-like proteins.

alanine and glycine side chains, respectively, that project outwards onto different sides of each polypeptide backbone and run along the *b* axis. When several chains are packed along the *c* axis, the alanine and glycine residues alternate as shown to produce a compact structure that is further stabilized by van der Waals forces between the amino acid side chains.

In general, it is not easy to characterize the hydrophobic macromolecules present in shell nacre because they tend to form an intractable organic goo at the bottom of the test-tube when extracted from the demineralized shell. However, some detailed data are now available for a few proteins isolated from nacre due to the onset of molecular cloning techniques in biomineralization (see Table 6.2 and further reading). In particular, two framework proteins from the nacreous layer of the Japanese pearl oyster, *Pinctada fucata*, have been recently studied. These are referred to as *MSI 60*, which is a glycine/alanine-rich silk fibroin-like macromolecule, and protein *N16*—the numbers refer to the molecular masses of 60 000 and 16 000, respectively. The complete amino acid sequence of N16 is now known. It turns out, however, that only protein MSI 60 is integral to the insoluble organic framework because N16 can be extracted as a soluble macromolecule when the hydrophobic matrix is treated with NH_4OH at pH 8.5. Indeed, N16 contains acidic amino acids in four regions of the macromolecule, although the repeating domains are $[\text{Asn-Gly}]_n$, which are polar rather than charged because they contain asparagine (Asn), which is an uncharged derivative of aspartate (Asp) (Fig. 6.22). Similar results were observed for a related protein, *N14*, extracted by similar methods from the nacreous layer of *Pinctada maxima*, which is a close (and presumably large) relative of the Japanese pearl oyster.

H H O
—N—C—C—
CH$_2$
C
O NH$_2$

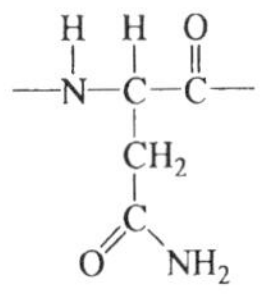

Asn

Fig. 6.22 Asparagine side chain.

In general, proteins N14 and N16 appear not to be acidic enough to suggest a role in calcium carbonate nucleation, nor sufficiently hydrophobic to act as a structural matrix alone. Instead, they may have several different functions at the interface between the silk fibroin-like proteins and the more highly acidic macromolecules that are present in the shell. For example, they could act as linker proteins that help to couple these components together.

New types of functions have also been suggested for another protein, referred to as *lustrin A*, which has been isolated from the insoluble matrix of the nacreous layer of the perforated ear-shaped shell of the red abalone (*Haliotis rufescens*). This protein is highly complex in its amino acid sequence and contains a modular structure built from multiple domains, many of which are similar to various other extracellular proteins, suggesting that lustrin A is a jack-of-all-trades when it comes to biomineralization. In particular, a domain close to the C-terminus has a sequence very similar to those found in *protease inhibitors*, which are a general family of macromolecules that prevent enzymes from degrading proteins. Thus, one novel function for lustrin A could be that it prevents the organic matrix from being attacked by enzymes that are present in the surrounding fluids.

Unfortunately, it is just as difficult to characterize the acidic macromolecules in nacre as it is the components of the insoluble hydrophobic framework. Although the macromolecules are soluble, they are often so highly negatively charged that they stick to every surface in sight. They are rich in carboxylate residues (typically 30 per cent aspartate (Asp) and 17 per cent glutamate (Glu)), serine residues with and without covalently linked phosphate groups, and sulfated polysaccharide side chains (see Figs 6.13 and 6.15 for molecular structures). Molecular cloning has provided some information on the repeat sequences present in a few of these macromolecules. For example, repeats of

- [Gly-Asp-Asn]
- [Gly-Glu-Asn]
- [Gly-Asn-Asn]

have been identified in *nacrein*, a polyanionic macromolecule extracted from the Japanese pearl oyster. There is circumstantial evidence that many of these macromolecules, like their hydrophobic counterparts, contain localized domains with the antiparallel β-pleated sheet secondary structure.

Another water-soluble macromolecule, *N66*—no prizes for guessing its molecular mass—has also been studied. An intriguing aspect of the amino acid sequence is that part of the molecule resembles a domain present in *carbonic anhydrase*, which, as we mentioned in Chapter 5 (Sections 5.2 and 5.4.2), is an enzyme involved in the interconversion of CO_2 and HCO_3^-. This suggests that some of the matrix macromolecules might function in the chemical control of supersaturation levels associated with the organic surface.

The organic matrix of shell nacre consists of structural and functional macromolecules. Structural macromolecules consist of antiparallel β-pleated sheet silk fibroin-like hydrophobic proteins, such as MSI 60, and are rich in alanine and glycine. In contrast, highly acidic glycoproteins, such as nacrein, are rich in aspartate, glutamate and serine, and bind to the hydrophobic scaffold where they probably function in aragonite nucleation. Other macromolecules (N14, N16, N66, lustrin A) are intermediate in character and are probably multifunctional. β-Chitin fibrils are sometimes sandwiched between the hydrophobic proteins.

6.6 Macromolecules and silica biomineralization—diatoms and sponges

Except in plants, where the deposition of silica generally occurs in the presence of an extracellular organic matrix such as cellulose (Fig. 6.19), most biogenic silicas are formed within intracellular vesicles. In most of these systems, no internal organic framework is apparent and the mineral phase is contained within the vesicle membrane, which acts as an external wall. In principle, the organic casing can be removed by partial degradation using strong acids without dissolution of the silica phase to give a range of matrix components analogous to the hydrophobic framework macromolecules of bones and shells. Occasionally, a distinct internal organic framework is associated with the silica biomineral, for example in silica sponge spicules (see below) and in certain freshwater algae such as *Synura petersenii* (see also Chapter 7, Section 7.3.1). The macromolecules associated with these internal matrices can only be isolated by dissolution of the silica structure, which is achieved using aqueous hydrofluoric acid. Because of their close association with the mineral phase, the released organic components will contain both structural and functional macromolecules (Table 6.2), analogous to the framework and acidic macromolecules, respectively, that are associated with calcium-containing biominerals. In general, therefore, the organic matrix associated with silica structures in biology can be rationalized by the same two-component structure–function model described in Section 6.2.

The silica shell (*frustule*) of *diatoms* has been extensively studied in recent years (see further reading). By judicious chemical treatment of the mineralized structure, three distinct types of proteins have been identified. These are:

- *frustulins*
- *HEPs*
- *silaffins*.

Glycoproteins termed *frustulins* are extracted from the frustule under relatively mild conditions using treatment with EDTA. This removes some of the proteins associated with the organic casing—particularly those that are stabilized by Ca^{2+} ion cross-links—without dissolving the mineral. The frustulins have relatively large molecular masses of about 75 000, and a modular structure. This includes an acidic domain which contains multiple repeats of a sequence consisting of cysteine (Cys) (Fig. 6.23), aspartate (Asp) and glutamate (Glu) (Fig. 6.13), and glycine (Gly) (Fig. 6.20) residues, with the conserved structure

$$[\text{Cys-Glu-Gly-Asp-Cys-Asp}]_n$$

H H O
—N—C—C—
CH$_2$
SH

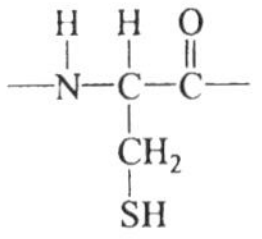

Cys

Fig. 6.23 Cysteine side chain.

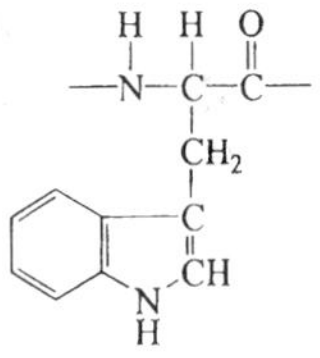

There are also domains enriched in proline (Pro) (Fig. 6.16), and hydrophobic regions consisting of polyglycine ($[\text{Gly}]_n$), or tryptophan (Trp) (Fig. 6.24). The combination of high molecular mass and hydrophobic residues suggests that the frustulins are framework macromolecules that play an important structural role in the organic casing surrounding the silica shell.

H H O
—N—C—C—
CH$_2$
C
CH
N
H

Trp

Fig. 6.24 Tryptophan side chain.

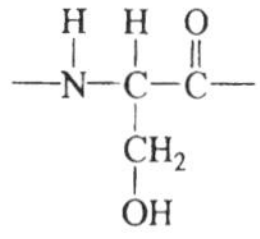

A second class of proteins, referred to as *HEPs*, or *HF-extractable proteins*, are associated strongly with the diatom frustule and can only be isolated if the silica shell is dissolved by hydrofluoric acid. These macromolecules, such as HEP200, are enriched in the hydroxy-functionalized amino acids serine (Ser) and threonine (Thr) (Fig. 6.25). Some of the extracted proteins contain about 25 mol% of serine/threonine, 25 mol% glycine and 20 mol% of acidic residues. The high number of polar residues suggests that these proteins are important functionally in the silicification process.

Another family of proteins with extremely high affinity for the mineral phase—hence the name, *silaffins*—is found in association with the HEP macromolecules when the silica frustule is dissolved. The silaffins are very different, however. They have a low molecular mass, which extends from 4000 to 17 000, and an unusual *polycationic* structure. The high positive charge arises from the regular spacing of clusters of lysine (Lys) and arginine (Arg) residues (Fig. 6.26) that are separated by serine and threonine residues along the polypeptide chain. Some of the lysine residues are modified with ε-*N*,*N*-dimethyl (DMLys) or 6 to 11 repeats of oligo-*N*-methyl-propylamine (MPALys) side chains (Fig. 6.26), which increase the number of positive

H H O
—N—C—C—
CH$_2$
OH

Ser

H H O
—N—C—C—
HC—OH
CH$_3$

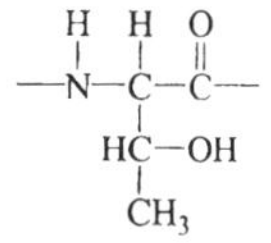

Thr

Fig. 6.25 Hydroxy amino acid side chains associated with HEP macromolecules.

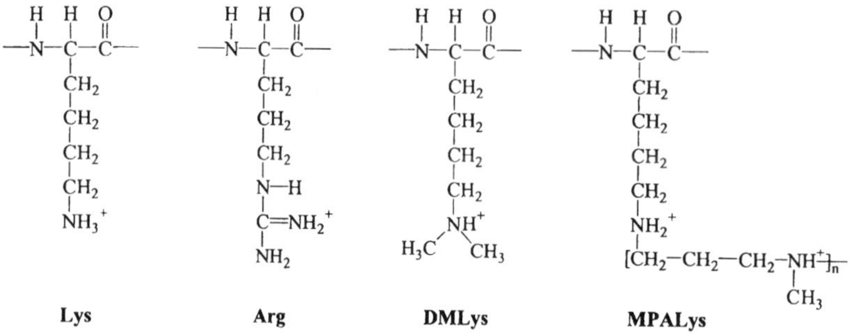

Lys **Arg** **DMLys** **MPALys**

Fig. 6.26 Cationic amino acid side chains found in silaffins.

charges associated with the silaffins. The highly polycationic nature of these proteins is consistent with their strong affinity for silica, and also suggests that the macromolecules are involved in the promotion of silicic acid polycondensation and intravesicular deposition of silica.

Silica *sponge spicules* are also formed in vesicles but unlike the diatom frustule they contain a well-defined and spatially separated internal organic matrix that can be easily extracted from the membrane components of the vesicle wall (see further reading). This is because the matrix consists of a millimetre-long axial filament that is occluded in the centre of the silica fibre. Dissolution of the mineral phase leaves an intact organic filament that is composed of a crystalline array of three types of protein subunits referred to as *silicatein* α, β and γ. The amino acid composition is somewhat similar to some of the HEP proteins extracted from the diatom shell wall, with 20 mol% serine/threonine, 15 mol% glycine and 20 mol% acidic amino acids. Interestingly, silicatein α has an amino acid composition similar to a proteolytic enzyme called *cathepsin* that is involved in hydrolysis reactions. As the polycondensation of silicic acid is effectively a 'reverse hydrolysis' reaction (water is produced during silica deposition), then the silicateins might be involved in controlling the rate and extent of mineralization of the silica spicules.

Proteins associated with the formation of the diatom frustule include organic casing components such as frustulins, and intra-silica macromolecules, enriched in serine and threonine residues (HEP proteins) or lysine and arginine residues (silaffins). The latter are highly cationic and contain unusual side-chain modifications. Sponge spicules, in contrast, contain a central organic filament constructed from the self-assembly of three different types of a serine/threonine-rich protein, silicatein.

6.7 Organic matrix-mediated nucleation

Having described some of the compositional and structural features associated with organic matrices in biomineralization, we now turn to the fundamental question concerning the role of these insoluble frameworks in controlling the nucleation of inorganic minerals such as calcium carbonate and calcium phosphate. Although our attention is focused principally on the surface chemistry of macromolecular surfaces that consist of functional groups such as charged amino acids and anionic sugar moieties, it is important to note that phospholipids (see Chapter 5, Section 5.1.1) can also play an important role in the control of nucleation. For example, the initial stage of cartilage calcification occurs in association with vesicles that have bilayer membranes enriched in a negatively charged phospholipid, *phosphatidyl serine* (Fig. 6.27). These lipids form stable complexes with Ca^{2+} ions that in turn attract HPO_4^{2-} anions with the consequence that nuclei are readily formed along the internal surface of the vesicle. Similar phospholipid–calcium-phosphate complexes are present in the cell wall membrane of certain bacteria that cause the adventitious precipitation of hydroxyapatite on tooth surfaces. See further reading for more details.

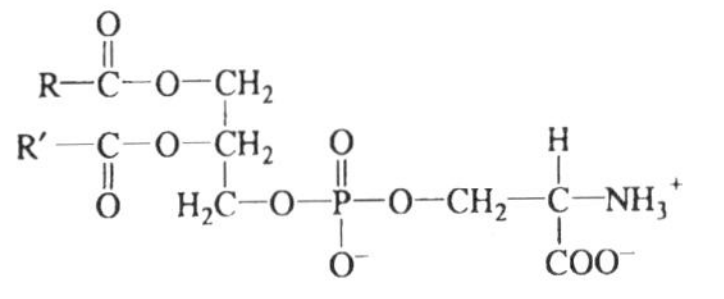

Phosphatidyl serine

Fig. 6.27 Phosphatidyl serine. R and R′ are long-chain moieties.

Whether made of phospholipids, polypeptides or polysaccharides, the central role of the organic matrix in controlling inorganic nucleation is *to lower the activation energy by reducing the interfacial energy.* This is illustrated in Fig. 6.28, where both the activation energy of nucleation ($\Delta G^*_{N(2)}$) and the size of the critical cluster required for nucleation ($r^*_{(2)}$) in the presence of the organic matrix (curve 2) are significantly reduced compared with the corresponding values ($\Delta G^*_{N(1)}$ and $r^*_{(1)}$) in the absence of the organic surface (curve 1). Several effects are associated with this mechanism:

- changes in the rate of nucleation
- site-specific organization of nucleation sites on the organic surface
- structural selectivity of mineral polymorphs
- crystallographic alignment of nuclei on the organic surface.

We showed in Chapter 4, Section 4.4, that the rate of nucleation is directly related to the activation energy required to produce the new solid–liquid interface associated with nuclei forming in the supersaturated solution. Thus, by lowering the activation energy through specific molecular interactions at certain positions on the organic matrix, both the rate of mineral nucleation and the site of inorganic deposition can be highly regulated. Moreover, by lowering the activation energy specifically for one polymorph rather than another, nucleation on the organic matrix could lead to the preferential deposition of one particular mineral structure. Changes in the surface structure

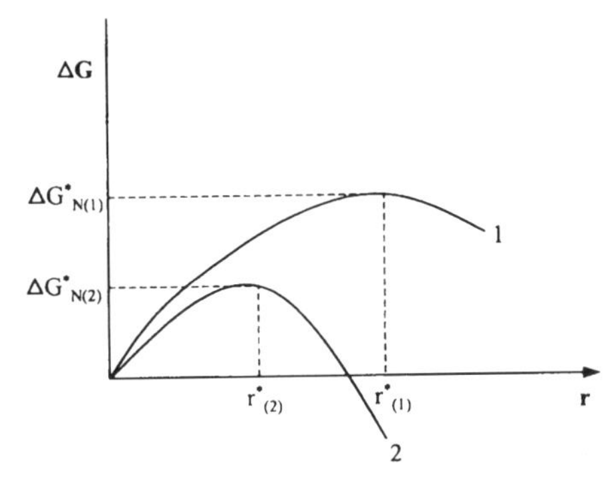

Fig. 6.28 Free energy curves for nucleation in the absence (1) and presence (2) of an organic surface.

and composition of the organic matrix might then result in the switching of polymorphs, and this might, for example, explain why aragonite and calcite are nucleated in the inner and outer layers, respectively, of certain seashells.

Figure 6.29 shows diagrammatically how the activation energies can be influenced in the presence of an organic matrix for the *structural control of nucleation* in biomineralization. If we consider a mineral nucleated in the absence (state 1) or presence (state 2) of an organic matrix, then the activation energy corresponding to state 2 is always lower than state 1 because the organic surface promotes nucleation. Now, the inorganic phase also has two possible polymorphs, A and B, where A is the more kinetically favoured in the absence of the organic matrix. Then, depending on the relative changes in the nucleation activation energies for A and B in the presence of the organic matrix, three possible outcomes can take place as shown in Fig. 6.29A, B and C, respectively:

- *promotion of non-specific nucleation* in which both polymorphs have reduced activation energies because of the presence of the matrix surface but there is no change in the outcome of mineralization.
- *promotion of structure-specific nucleation* of polymorph B due to a more favourable crystallographic recognition at the matrix surface. Thus, the activation energy of state 2 for structure B falls lower than that of state 2, structure A.
- *promotion of a sequence of structurally non-specific to highly specific nucleation* depending on how the levels of recognition of nuclei A and B and the reproducibility of matrix structure change with external factors such as genetic, metabolic and environmental processes.

The process of organic matrix-mediated nucleation becomes even more remarkable when one considers that not only is the structure of the biomineral controlled by the organic surface but also in certain cases the nuclei are crystallographically aligned with regard to the underlying matrix. For example, the aragonitic tablets in the nacreous layers of the shell are preferentially nucleated so that the *c* axis of the unit cell is perpendicular to the plane of the organic sheets (see Fig. 6.18). In enamel, the interwoven crystalline rods consist of bundles of long hydroxyapatite crystals co-aligned with

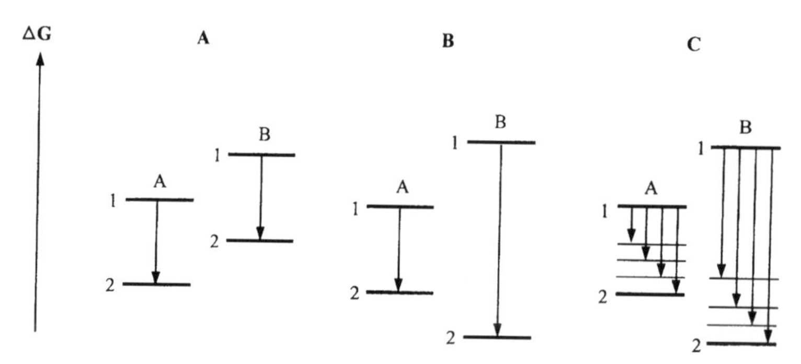

Fig. 6.29 Structural control by organic matrix-mediated nucleation. See text for details.

their c axes perpendicular to the β-sheets of the surrounding enamelin proteins. And in bone, the plate-like hydroxyapatite crystals are oriented such that the c axis is aligned with the collagen axis (see Fig. 6.12).

Sometimes it is possible to explain the crystallographic alignment of certain biominerals by simple physical mechanisms. For example, in Chapter 5, Section 5.1.4, we showed that calcite crystals in the avian eggshell become preferentially oriented along their c axes by physical limitations placed on the growth of spherulites originating from a closely spaced set of nucleation sites. But in most systems oriented nucleation is considered to arise from specific molecular mechanisms that lower the activation energy of nucleation along a particular crystallographic direction. The situation is therefore very similar to that described in Fig. 6.29 for the control of polymorph structure, except that the structural states represented by A and B now correspond to two different crystal faces. For example, in the cubic system these could have Miller indices such as {100} and {111}. Following the previous discussion, nucleation of the less stable {111} could then occur if appropriate interfacial interactions specifically lowered the activation energy for this face rather than the more stable {100} surface.

The types of molecular interactions that can account for these interfacial processes of nucleation are discussed in the following sections of this chapter.

A central tenet of organic matrix-mediated nucleation in biomineralization is that the activation energy for inorganic nucleation is lowered by specific interfacial interactions between functional groups on the macromolecular surface and ions in the supersaturated solution. This can lead to control over the rate of nucleation, the number and organization of nucleation sites, polymorph selectivity and oriented nucleation.

6.7.1 Interfacial molecular recognition

The specific lowering of the activation energy for nucleation in organic matrix-mediated biomineralization is thought to involve a number of different types of interfacial interactions between ions of the mineral phase and functional groups on the macromolecular surface (Fig. 6.30). In general, these interactions involve some form of *molecular recognition* at the inorganic–organic interface. For example, matching of charge, polarity, structure and stereochemistry can give rise to specific changes in the activation energy that can be fine-tuned for controlling the nucleation rate, site-specificity, mineral structure and crystallographic alignment.

In many respects, the surface of the organic matrix is analogous to an enzyme in solution, with the inorganic nucleus taking the place of the substrate molecule that usually binds to the enzyme. In both cases, the lowering of the activation energies for nucleation and reaction, respectively, depends on *shape (topography) and chemical complementarity* between the two components acting over molecular distances. But nucleation differs in that long-range interactions arising from the electrostatic forces and periodic structures of ionic surfaces also need to be considered.

As we shall discuss below, most theories of interfacial molecular recognition are based on the collective ordering of individual ions on the organic

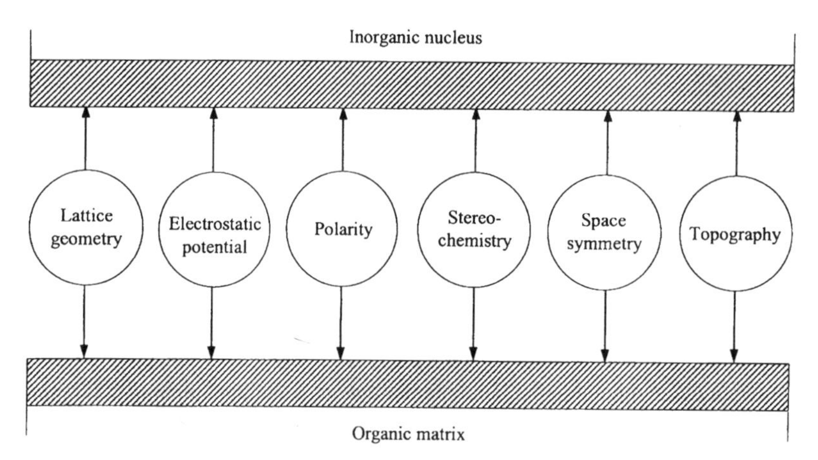

Fig. 6.30 Molecular recognition at inorganic–organc interfaces in biomineralization.

surface and the influence of binding sites and their long-range organization on nucleation. However, an alternative description involving the preferential organic matrix-mediated selection and stabilization of nuclei already present in the supersaturated solution is feasible. In this case, interfacial recognition is at the supramolecular rather than molecular level. However, the mechanistic differences between these models is not as large as one might initially expect because both involve recognition processes based on charge, structural and stereochemical matching.

Interfacial molecular recognition is a key concept in organic matrix-mediated biomineralization. Lowering of the activation energy for nucleation is considered to arise from the matching of charge, polarity, structure and stereochemistry at the interface between an inorganic nucleus and an organic macromolecular surface. The shape of the interface and the degree of chemical complementarity are important factors in this process.

6.7.2 Electrostatic accumulation—the ionotropic model

The most fundamental property of the organic matrix in relation to the nucleation of an inorganic mineral concerns the distribution of charges across its surface. This gives rise to a simple form of interfacial complementarity in which the charge distribution between anions and cations in planes of the mineral lattice is mimicked by the surface arrangement of ions bound to organic ligands exposed at the organic surface. The activation energy of nucleation is then lowered because the ions are already organized at least in two dimensions by the organic matrix.

We might therefore expect to find a general correspondence between the types of functional groups present on the surface of the organic matrix and the chemical nature of the mineral phase. To a limited extent this appears to be true. For example, we mentioned earlier in Sections 6.3.2, 6.4 and 6.5 that calcium-containing biominerals are often associated with highly acidic macromolecules that are enriched in carboxylate (aspartate, glutamate) and phosphorylated (phosphoserine) amino acids, and sulfated sugar groups.

These anionic ligands form reasonably strong electrostatic bonds with Ca^{2+} ions so that the metal cations are accumulated at specific sites on the organic surface. Similarly, silica biominerals are formed in the presence of hydroxy-rich macromolecules, such as polysaccharides and proteins enriched in serine and threonine amino acids, or positively charged proteins such as *silaffins* that contain cationic domains of modified lysine residues (Section 6.6). Again, the general principle of interfacial complementarity applies, as both polar (-OH) and electrostatic interactions are known to be important in controlling the deposition of silicate precursors from aqueous solution.

To this simple picture of organic matrix-induced ion binding we need to add a mechanism for increasing the ordering process so that it extends over distances large enough to accommodate the size of a typical inorganic nucleus—2 or 3 nm perhaps. One possibility is that the binding sites are clustered together to produce localized regions of *high spatial charge density* on the organic matrix. For example, Fig. 6.31A shows a localized molecular groove or pocket containing a large number of closely packed charged residues initially filled with water molecules. These sites accumulate and stabilize sufficient numbers of ions in the concave space such that nucleation of the biomineral occurs (Fig. 6.31B). This mechanism is sometimes referred to as *ionotropy* (see further reading). It can be considered as a method for increasing the local supersaturation at the organic surface or as a means for lowering the nucleation activation energy, or both. In the latter case, the structural rearrangement of the initially dispersed highly solvated ionic cluster into the more ordered structure of the condensed phase (Fig. 6.31C) is dependent on the degree of dynamic freedom imposed by the matrix interactions. If the cluster is going to click into place as an ordered lattice then the ions need to have some mobility and solvent molecules have to be expelled. The binding therefore needs to be strong enough to accumulate large numbers of ions but not too strong because this would fix the ions to organic ligands on the matrix surface. In general, it is thought that *high capacity, low affinity* interactions are required to facilitate the movement of ions into the periodic lattice sites of a crystalline nucleus.

Another possibility is that instead of individual hydrated ions, preformed nuclei and ionic clusters present in the supersaturated solution are stabilized at the organic surface by a *charge-matching* mechanism. In this case, tiny

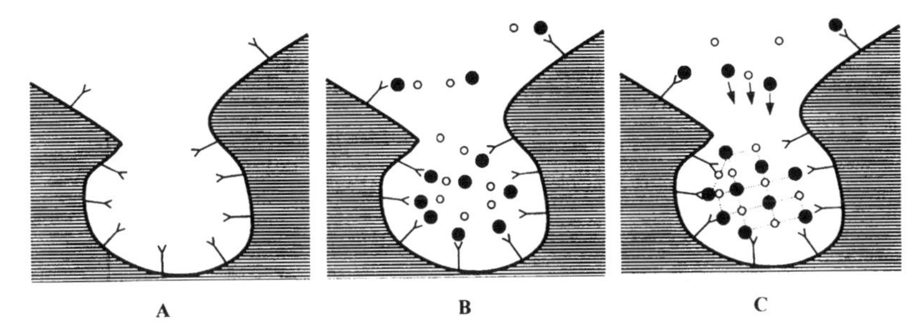

Fig. 6.31 Ionotropic mechanism for inorganic nucleation on organic surfaces. See text for details.

nuclei migrate across the organic matrix to energetically favourable sites with matching spatial charge density. Once they are pinned down onto the organic surface, the clusters undergo internal crystallization due to the lowering of their surface energy by association with the organic matrix.

Ionotropy—the localized accumulation of ionic charge by high capacity, low affinity binding in regions of high spatial density—is a potential mechanism for lowering the activation energy of inorganic nucleation on functionalized organic surfaces.

6.7.3 Nucleation in ferritin

The nucleation of iron oxide nanoparticles within the protein cavity of ferritin illustrates the importance of spatial charge and electrostatic accumulation in controlling the site-specific deposition of a biomineral. The structure and self-assembly of ferritin were described in Chapter 5, Section 5.1.2. We now focus our attention on the mechanism that leads to preferential nucleation within the 8-nm-diameter polypeptide cage.

A lot of information is available because of the remarkable ability of ferritin to mineralize in the test-tube. The experiment involves the removal of the native iron oxide (ferrihydrite) cores from the protein by reductive dissolution using an agent such as thioglycolic acid or sodium dithionite. This results in intact but empty protein molecules, referred to as *apoferritin*, which can then be reconstituted with iron oxide simply by adding low concentrations of Fe^{II} ions to the buffered solution at pH 6.5 in air. Amazingly, the solution turns red–brown in colour but no precipitation is observed because all the iron oxide ends up inside the ferritin molecules so the mineral remains dispersed in solution in the form of polypeptide-encapsulated nanoparticles. This simple but elegant experiment allows a detailed study of the kinetics (see further reading), which at low levels of Fe^{II} (< 30 Fe^{II} per protein molecule) take place according to the reaction

$$2Fe^{II} + O_2 + 4H_2O \rightarrow 2FeOOH + H_2O_2 + 4H^+$$

At higher loadings (> 250 Fe^{II} per protein molecule), however, the stoichiometry changes to

$$4Fe^{II} + O_2 + 6H_2O \rightarrow 4FeOOH + 8H^+$$

By changing selective amino acids in the protein using a process called *site-directed mutagenesis*, the role of specific regions of the protein in nucleation can be determined. The current level of understanding is as follows (see further reading for more details). It is clear that nucleation of the Fe^{III} oxide nanoparticles in the cavity is preceded by aerial oxidation. It turns out that these two processes occur at different sites in the protein molecule (Fig. 6.32). Oxidation takes place by a catalytic step in which Fe^{II} ions are bound to a specific site in the protein for reaction with molecular dioxygen. This site is referred to as the *ferroxidase centre* and is present only in the H-chain subunit. The site is located within the centre of the four-helix bundle of the subunit and involves a single Fe^{II} ion bound to several amino acid residues, including three glutamic acids (Glu) at positions 27, 62 and 107

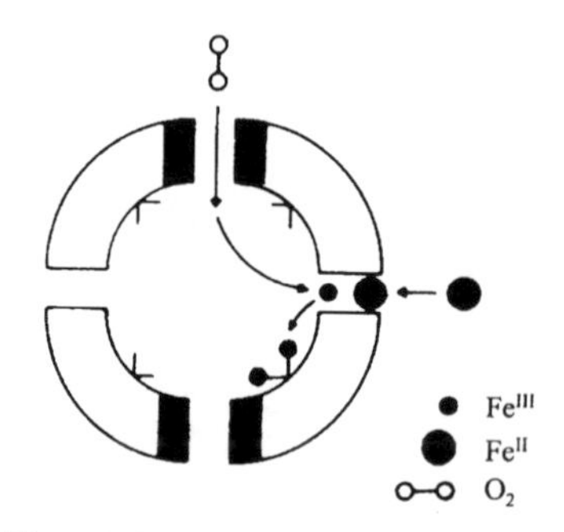

Fig. 6.32 Two-site mechanism for mineralization in ferritin.

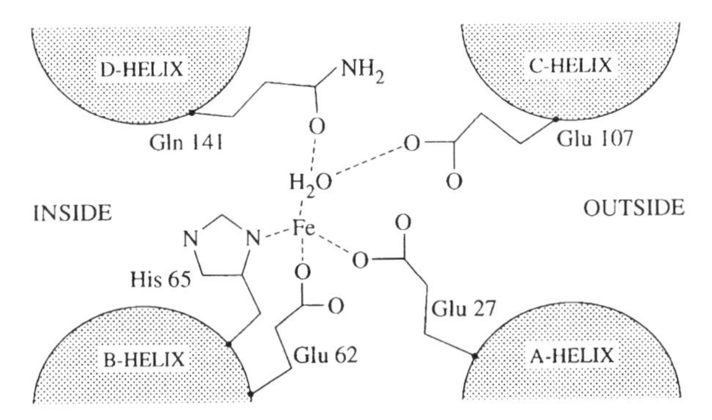

Fig. 6.33 Binding of Fe^{II} at the ferroxidase centre of ferritin.

along the polypeptide chain, a glutamine residue (Gln) at position 141, and a histidine side chain at residue 65 (Fig. 6.33).

Although oxidation of Fe^{II} to Fe^{III} proceeds at the ferroxidase centre, nucleation of ferrihydrite occurs on the inner surface of the polypeptide cavity. The working hypothesis is that Fe^{III} ions formed at the ferroxidase centre migrate into the cavity because of the large electrostatic field of three neighbouring glutamate residues (Glu61, 64, 67) located 0.7 nm away on the cavity surface (Fig. 6.34). The glutamates are arranged along a groove at the interface of two subunits so there is a total of six negatively charged groups in close proximity. This localized anionic patch has sufficient spatial charge density to induce the accumulation of cationic Fe^{III} species, and lower the activation energy of nucleation. Because the hydrated Fe^{III} complexes are now clustered together, they become irreversibly cross-linked by condensation reactions involving the formation of oxo (Fe–O–Fe) and hydroxy (Fe–OH–Fe) bridges (Fig. 6.35).

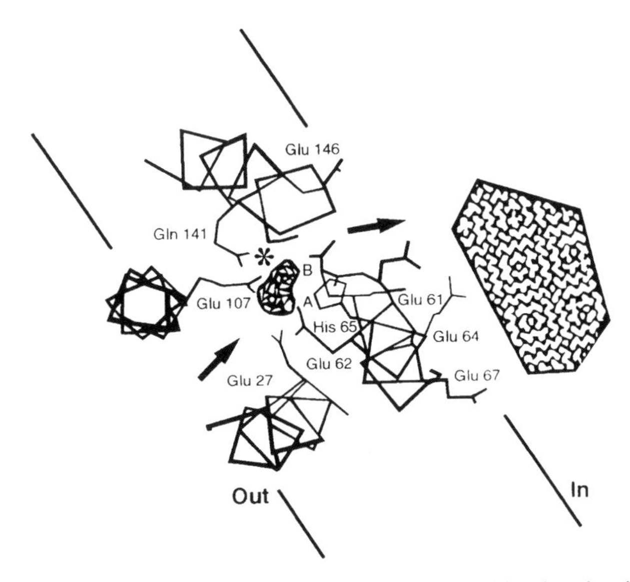

Fig. 6.34 Cross-section of the polypeptide shell of ferritin showing buried ferroxidase centre (AB) and nucleation site on internal surface. Arrows denote steps in the mineralization mechanism.

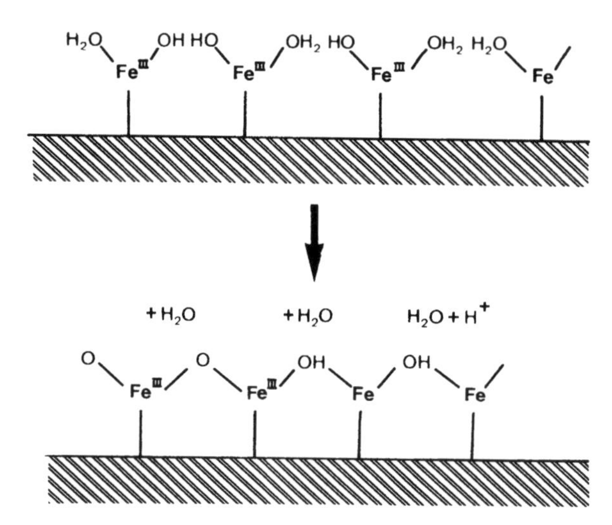

Fig. 6.35 Polycondensation and cross-linking of Fe^{III} ions bound at the nucleation site in ferritin.

Once formed, the iron oxide clusters grow by continual addition of Fe^{III} species to the existing surface such that in time all the added Fe^{II} is oxidized and precipitated specifically within the protein cavity.

Nucleation of Fe^{III} oxide in ferritin is preceded by Fe^{II} oxidation at the ferroxidase centres, and involves the electrostatic accumulation of Fe^{III} species at an anionic patch of glutamate residues located on the internal surface of the polypeptide cage.

6.7.4 Surface topography

In ferritin nucleation, the clustering of the glutamic acid residues along a molecular groove at the interface of two polypeptide subunits concentrates the anionic charge within a relatively small area of the surface. This example illustrates how charge accumulation along an organic matrix can be influenced on the nanometre length scale by the local *surface topography* of the nucleation site. Changing the localized shape of the surface affects the spatial charge distribution of functional groups and hence their ability to stabilize the formation of inorganic clusters from supersaturated solutions. There are three main possibilities, shown in Fig. 6.36A, B and C, respectively:

- *concave* surfaces—like ferritin, they give rise to a high spatial charge density and three-dimensional clustering of ions, and are good nucleation sites.
- *convex* surfaces—dissipate the charge density and so are unlikely to act as nucleation sites but can be used to limit the number of nucleation centres across the surface of the organic matrix.
- *planar* substrates—have localized charge distributions and can act as two-dimensional nucleation sites.

The case of planar substrates is particularly interesting because, in principle, regular arrays of functional groups can be distributed across the surface and

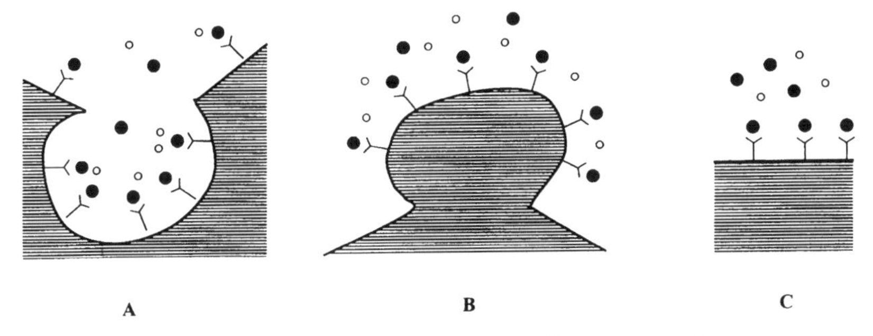

Fig. 6.36 Effect of surface topography on the distribution of ionic charge on organic surfaces: (A) concave; (B) convex; (C) planar interfaces.

used to template a periodic arrangement of surface-bound ions on the organic matrix. This is discussed in more detail in the following section.

Electrostatic accumulation of ions on organic surfaces is influenced by the localized clustering of ligands and their spatial charge distribution, which in turn depend on the surface structure and topography of the organic matrix

6.7.5 Structural matching—the geometric model

By binding ions in a regular array at the surface of an organic matrix, the lowering of the interfacial energy associated with nucleation can be coupled with *structural control* so that the resulting nucleus is oriented along a preferred crystallographic direction. This concept is based on the idea of *geometric or structural matching* (*epitaxy*) that we discussed in Chapter 4, Section 4.5. With respect to biomineralization, the close matching arises between lattice spacings in certain crystal faces and distances that separate functional groups periodically arranged across the organic surface. For example, in Fig. 6.37, anionic residues are spaced at regular intervals along the organic matrix with a repeat distance that is commensurate with a lattice spacing x that lies within the top face but not spacing y of the side face. Because crystal faces are two-dimensional, geometric matching must occur along two different

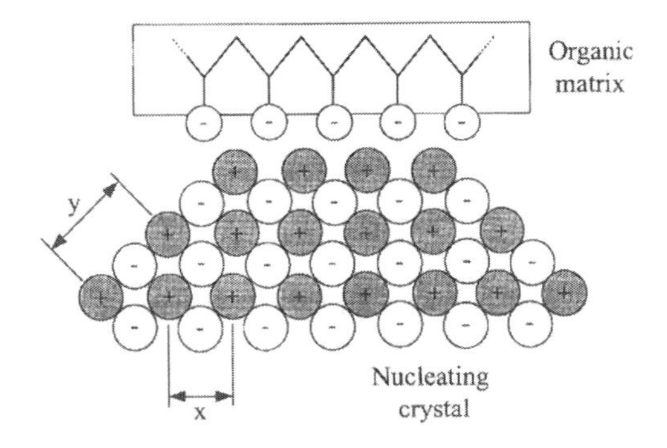

Fig. 6.37 Oriented nucleation by structural matching at an inorganic–organic interface.

crystallographic directions for oriented nucleation. This implies that the matrix has to have a crystalline surface that is locally flat.

Although the concept of structural matching is elegant and in fact rather popular amongst biomineralists, the experimental evidence is somewhat limited. Perhaps the best model we have is based on X-ray and electron diffraction studies of thin flakes of nacre from the mollusc shell (see further reading). The results show that the *a* and *b* axes of the antiparallel β-pleated sheet of the matrix are aligned with the *a* and *b* crystallographic directions of the aragonite lattice. That is, the crystals are oriented such that the (001) crystal face (the *ab* plane) and *c* axis are parallel and perpendicular to the underlying matrix surface, respectively. As shown in Fig. 6.38, if we now compare the distances between Ca^{2+} ions in the (001) face with the matrix periodicity then there is very good matching along the *a* axes (0.496 nm and 0.47 nm, respectively). The lattice match along the corresponding *b* axes is not as high (0.797 nm and 0.69 nm, respectively) but the periodicities are essentially commensurate over longer distances such as that spanning seven Ca^{2+} ions (4.8 nm).

But what controls the organization of the Ca^{2+} ions along the antiparallel β-pleated sheet? Well, we know from our discussions in Section 6.5 that shell proteins are highly acidic macromolecules, being enriched, for example, in aspartic acid (Asp) residues that can act as ligands for calcium binding. So if these are arranged along the hydrophobic matrix such that repeated domains of [Asp-X] (where X is a neutral amino acid) also adopt a β-sheet conformation, then there is a strong correlation between the spacings of the carboxylate groups and the theoretical lattice arrangement of Ca^{2+} ions in the (001) surface of aragonite (Fig. 6.39). The situation is complicated, however, by the need to balance the electrostatic interactions—one Ca^{2+} requires two aspartate ligands for example. The binding of calcium to carboxylate groups is therefore usually cooperative involving at least two or three ligands, so there needs to be enough flexibility in the geometric arrangement of the aspartate residues to accommodate these stereochemical requirements if the (001) face is to be preferentially nucleated.

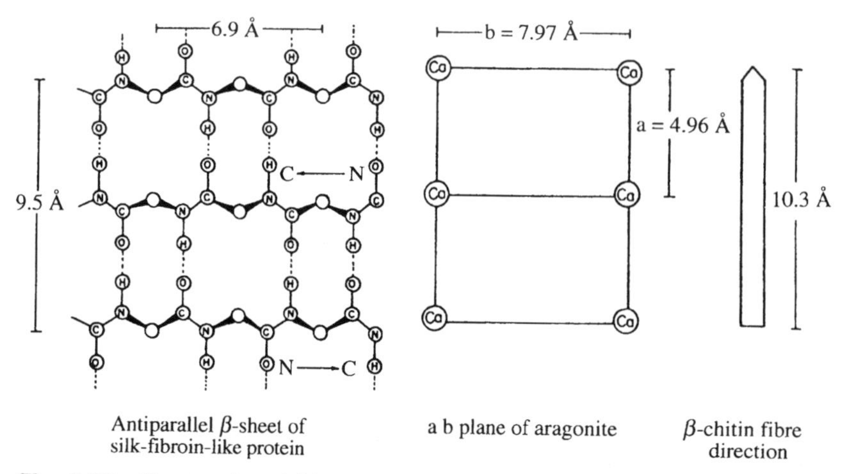

Fig. 6.38 Structural model for geometric matching in shell nacre.

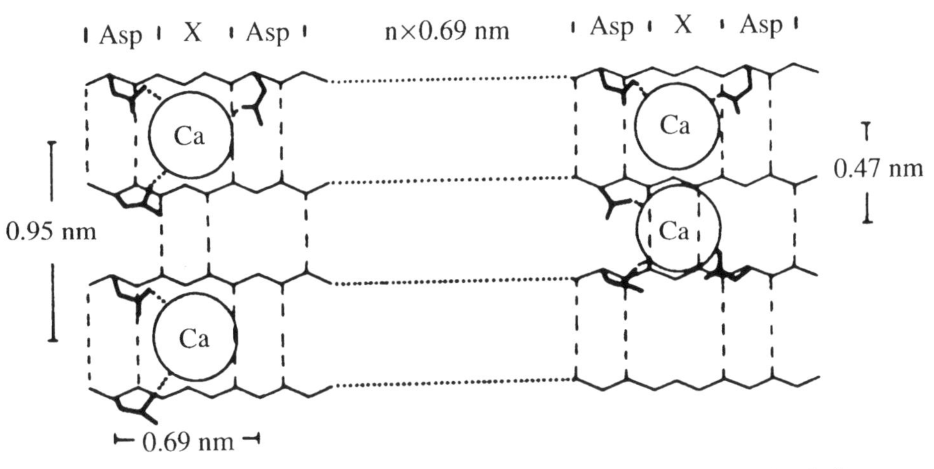

Fig. 6.39 Role of [Asp-X] domains in Ca^{2+} binding and oriented nucleation in shell nacre.

One drawback of this model is that it turns out that the binding constants for Ca^{2+}-carboxylate complexes are not particularly high, so it is questionable whether a periodic arrangement of Ca^{2+} would be sufficiently stable to sustain an active site for nucleation. However, if we look back to Section 6.5, we see that the shell proteins are glycosylated, that is they are modified with sugar-like residues. One possibility is that covalently bound sulfated polysaccharides act as 'antennae' that help to accumulate sufficient numbers of Ca^{2+} ions around the aspartate binding/nucleation sites on the antiparallel β-pleated sheet of the matrix. This model—see further reading for more details—therefore incorporates both ionotropic and epitaxial mechanisms to account for matrix-mediated oriented nucleation in biomineralization.

Structural (geometric) matching at inorganic–organic interfaces is a key concept in oriented nucleation in biomineralization. In this model, distances between regularly spaced binding sites on the surface of the organic matrix are commensurate only with certain lattice spacings in particular crystal faces. The model has been used to explain the specific nucleation of the (001) face of aragonite on the surface of highly acidic antiparallel β-pleated sheet proteins in shell nacre.

6.7.6 The stereochemical model

Even with the accommodation of ionotropic factors into the structural model of oriented nucleation, there are still some reservations. First, the epitaxial description works fine for inorganic crystals growing off the faces of other inorganic substrates with related structures, but macromolecular surfaces are generally not flat and well-ordered. Second, macromolecules often display considerable molecular dynamics in their surface residues as well as along the polypeptide or polysaccharide backbone so it is difficult to fit a constant distance to the ligand spacing. And third, there is the '*calcite–aragonite problem*'.

This problem has been known to keep some biomineralists awake at night. It goes like this (yawn!). The structural matching between aspartate residues spaced in tandem repeats of $[Asp\text{-}X]_n$ and Ca^{2+} ions in the (001) face of aragonite appears at first sight not only to be a very elegant mechanism for explaining the crystallographic orientation of the nacreous layer but also the selection of a different polymorph compared with the calcite-containing outer layer of the shell. But an inspection of the crystal structures shows that the distances between Ca^{2+} ions in the (001) face of calcite are almost identical to those in the (001) face of aragonite! The Ca^{2+} ions are arranged in the calcite (001) face in a hexagonal lattice with a uniform spacing of 0.499 nm, while in the aragonite (001) face there is a pseudo-hexagonal lattice of Ca^{2+} ions spaced at distances of either 0.496 or 0.797 nm along the orthogonal *a* and *b* axes, respectively (Fig. 6.40). Superimposing the lattices therefore gives a reasonable match over several unit cells of the (001) planes. So what aspects of the matrix discriminate in favour of the oriented nucleation of aragonite?

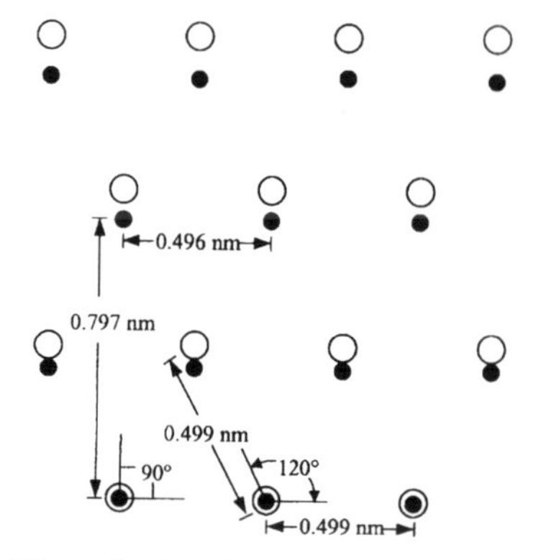

Fig. 6.40 Superimposition of calcite (open circles) and aragonite (filled circles) (001) crystal faces.

We might answer this by proposing a mechanism in which the structural control intrinsic to the matrix surface is only effective within a relatively narrow window of the solution conditions. For example, the ionic strength, level of supersaturation and amount of Mg^{2+} and other extraneous ions and soluble macromolecules present in the surrounding fluid could just give aragonite nucleation the edge over calcite. Another possibility is to look more closely at the aragonite and calcite structures. It turns out that although the distances between Ca^{2+} ions are very similar in the (001) faces of the two polymorphs, the arrangement of the CO_3^{2-} anions is different. That is, the *stereochemistry* around the Ca^{2+} ions is significantly changed. For example, the coordination numbers for Ca^{2+} ions in the calcite and aragonite lattices are six and nine respectively, so the arrangement and spacings of the oxygen atoms in the (001) faces of the polymorphs are defining aspects of the surface structure. In particular, although the CO_3^{2-} anions lie parallel to the (001) plane in both calcite and aragonite, those in calcite are all oriented in the same direction whereas there are two different arrangements for aragonite.

The question then arises whether the organization of the oxygen atoms of the aspartate side chains on the antiparallel β-pleated sheet corresponds more closely with the stereochemical requirements of the (001) face of aragonite than calcite. This would imply that when Ca^{2+} ions bind to the organic matrix the resulting coordination geometry created around each cation resembles the aragonite structure to such an extent that the interfacial energy associated with nucleation of the (001) face of aragonite is specifically lowered. As the aspartate ligands have only two oxygen atoms that can interact with the Ca^{2+} ions, whereas the CO_3^{2-} anions in the crystal structure have three, then the stereochemical matching will be somewhat imprecise but at its best where the crystal face contains calcium sites surrounded by bidentate ligands.

At the moment, there is no proof that stereochemical correspondence is responsible for the selection of calcium carbonate polymorphs in biomineralization. However, there are some positive indicators from crystallization experiments in model systems; see further reading for details.

The stereochemical correspondence between the coordination environment of ions in specific crystal faces and the arrangement of ligands around ions bound to the surface of an organic matrix is a potential factor in the oriented nucleation and selectivity of biomineral polymorphs, such as calcite and aragonite.

6.8 Summary

The evolution of organic matrix-mediated processes of biomineralization has transformed life's ability to assimilate hard inorganic materials into functional structures such as bones, shells and teeth. In this chapter we have described the main types of organic matrix and their important chemical role in controlling inorganic nucleation. A general model was presented in which the organic matrix consists of a structural framework of hydrophobic macromolecules with surface-anchored acidic macromolecules that act as a nucleation surface for biomineralization. The insoluble macromolecules are used to partition the extracellular space and provide a framework for mechanical support, while the acidic macromolecules interact with ions in the surrounding supersaturated solution and influence the activation energy barrier associated with nucleation.

The structural framework of the organic matrix originates from the cross-linking of specific types of building units, such as fibrils, nanospheres or sheets, which are in turn constructed from macromolecules with unique molecular and supramolecular structures. Collagen in bone, for example, consists of fibrils constructed from a staggered periodic arrangement of cross-linked triple-stranded tropocollagen helices, each of which contains non-coaxial helical polypeptides with repeat triplets, such as glycine–proline– hydroxyproline. Tiny plate-like crystals of hydroxyapatite nucleate within holes and grooves that are regularly spaced within the crystalline fibrils. In contrast, in enamel the hydroxyapatite crystals grow within a matrix of hydrophobic amelogenins that self-assemble into nanospheres, and which are progressively removed from the tissue during mineralization. A very different type of structural matrix is employed in shell nacre, in which antiparallel β-pleated sheet silk fibroin-like proteins, rich in alanine and glycine, are used to construct hydrophobic layers sometimes in association with chitin fibrils.

The acidic macromolecules associated with the above hydrophobic components vary considerably in detail but have similar generic properties. For example, the non-collagenous glycoproteins in bone are enriched in repeat sequences of anionic residues such as aspartate, glutamate and phosphoserine. Similar sequences are found in enamelins and certain shell proteins, which can sometimes adopt a β-pleated sheet secondary structure. Proteoglycans with anionic sulfate side chains are also present in large amounts in bone. Significantly, macromolecules associated with silica mineralization in diatoms are very different. They tend to be proteins enriched in polar serine and threonine residues or basic side chains such as lysine and arginine, all of which can interact strongly with the negatively charged and hydroxylated silica surface.

In the second half of this chapter we described how organic surfaces lower the activation energy for inorganic nucleation by molecular recognition involving the matching of charge, polarity, structure and stereochemistry at the matrix–nucleus interface. Accumulation of ionic charge at the matrix surface is influenced by the local clustering of ligands, which in turn is dependent on the surface structure and topography of the organic matrix. For example, nucleation of iron oxide in ferritin involves the electrostatic accumulation of Fe^{III} species at glutamate residues located in a groove that runs along part of the internal surface of the polypeptide cage. If the matrix ligands are spaced at regular distances then ion-binding can result in structural (geometric) matching at the inorganic–organic interface, which gives rise to oriented nucleation. Such a model has been proposed for the nucleation of the (001) face of aragonite on the surface of highly acidic antiparallel β-pleated sheet proteins in shell nacre, although stereochemical factors also need to be considered.

Throughout this chapter there has been a focus on *organic structure*, and its expression across a range of length scales. We have described the macromolecular organization of polypeptide and polysaccharide chains, supramolecular assemblies of multichain helices, nanospheres and protein–polysaccharide (proteoglycan) complexes, and microscopic architectures of cross-linked collagen fibrils and laminated organic sheets. In contrast, our discussion of mineral structure has been limited to the molecular level of unit cells and how these building blocks are influenced by interfacial recognition and oriented nucleation. We have said very little about how the mineral phase is influenced as the scale of organization increases. (In fact, such is the lure of the organic matrix that you can attend some scientific conferences, notably those on bone biomineralization, and hardly ever come across a serious mention of calcium phosphate!)

In the next two chapters, therefore, we take a look at how inorganic minerals adapt to the longer range force fields of biological organization. There are two intriguing outcomes—life-like inorganic morphologies (Chapter 7) and hierarchical multicomponent architectures (Chapter 8)—both of which highlight fascinating lessons for new areas of materials chemistry as described in Chapter 9.

Further reading

Addadi, L. and Weiner, S. (1985). Interactions between acidic proteins and crystals: stereochemical requirements in biomineralization. *Proc. Natl. Acad. Sci. U.S.A.*, **82**, 4110–4114.

Addadi, L. and Weiner, S. (1989). Stereochemical and structural relations between macromolecules and crystals in biomineralization. In *Biomineralization: chemical and biochemical perspectives* (ed. Mann, S., Webb, J. and Williams, R. J. P.), pp. 132–156. VCH Verlagsgesellschaft, Weinheim.

Addadi, L., Moradian, J., Shay, E., Maroudas, N. G. and Weiner, S. (1987). A chemical model for the cooperation of sulfates and carboxylates in calcite crystal nucleation: relevance to biomineralization. *Proc. Natl. Acad. Sci. U.S.A.*, **84**, 2732–2736.

Boskay, A. L. (1986). Phospholipids and calcification: an overview. In *Cell mediated calcification and matrix vesicles* (ed. Ali, S. Y.) pp. 175–179. Elsevier, Amsterdam.

Currey, J. (1984). *The mechanical adaptations of bones*. Princeton, NJ.

Currey, J. (1984). Effects of differences in mineralization on the mechanical properties of bone. *Proc. R. Soc. London B*, **304**, 509–518.

Deutsch, D., Palmon, A. Fisher, L. W., Kolodny, N., Termine, J. D. and Young, M. F. (1991). Sequencing of bovine enamelin (tuftelin): a novel acidic enamel protein. *J. Biol. Chem.*, **266**, 16021–16028.

Fincham, A. G. and Simmer, J. P. (1997). Amelogenin proteins of developing dental enamel. In *Dental enamel* (Ciba Foundation Symposium; 205), pp. 118–134. Wiley, Chichester.

Fischer, L. W. (1985). The nature of the proteoglycans of bone. In *The chemistry and biology of mineralized tissues* (ed. Butler, W. T.), pp. 188–195. EBSCO Media, Birmingham, AL.

Greenfield, E. M., Wilson, D. C. and Crenshaw, M. A. (1984). Ionotropic nucleation of calcium carbonate by molluscan matrix. *Am. Zool.*, **24**, 925–932.

Harrison, P. M., Andrews, S. C., Artymiuk, P. J., Ford, G. C., Guest, J. R., Hirzmann, J. *et al.* (1991). Probing structure–function relations in ferritin and bacterioferritin. *Adv. Inorg. Chem.*, **36**, 449–486.

Hecky, R. E., Moppr, K., Kilham, P. and Degens, E. T. (1973). The amino acid and sugar composition of diatom cell walls. *Mar. Biol.*, **19**, 323–331.

Heywood, B. R. and Mann, S. (1994). Template-directed nucleation and growth of inorganic materials. *Adv. Mater.*, **6**, 9–20.

Katz, E. P. and Li, S. (1973). The intermolecular space of reconstituted collagen fibrils. *J. Mol. Biol.*, **21**, 149–158.

Kono, M., Hayashi, N. and Samata, T. (2000). Molecular mechanism on the nacreous layer formation in *Pinctada maxima*. *Biochem. Biophys. Res. Commun.*, **269**, 213–218.

Kröger, N., Deutzmann, R. and Sumper, M. (1999). Polycationic peptides from diatom biosilica that direct silica nanosphere formation. *Science*, **286**, 1129–1132.

Miller, A. (1984). Collagen: the organic matrix of bone. *Philos. Trans. R. Soc. London B*, **304**, 455–477.

Parkes, E. W. (1965). *Braced frameworks*. Pergamon Press, Oxford.

Perry, C. C. and Keeling-Tucker, T. (2000). Biosilicification: the role of the organic matrix in structure control. *J. Biol. Inorg. Chem.*, **5**, 537–550.

Samata, T., Hayashi, N., Kono, M., Hasegawa, K., Horita, C. and Akera, S. (1999). A new matrix protein family related to the nacreous layer formation of *Pinctada fucata*. *FEBS Lett.*, **462**, 225–229.

Shen, X., Belcher, A. M., Hansma, P. K., Stucky, G. D. and Morse, D. E. (1997). Molecular cloning and characterization of lustrin A, a matrix protein from shell and pearl nacre of *Haliotis rufescens*. *J. Biol. Chem.*, **272**, 32472–32481.

Shimizu, K., Cha, J., Stucky, G. D. and Morse, D. E. (1998). Silicatein α: cathepsin L-like protein in sponge biosilica. *Proc. Natl. Acad. Sci. U.S.A.*, **95**, 6234–6238.

Stryer, L. (1988). *Biochemistry*. W. H. Freeman, New York.

Sun, S. and Chasteen, N. D. (1992). Ferroxidase kinetics of horse spleen apoferritin. *J. Biol. Chem.*, **267**, 25160–25166.

Termine, J. D. (1986). Bone proteins and mineralization. *Rheumatology*, **10**, 499–509.

van der Wal, P. (1989). Structural and material design of mature mineralized radular teeth of *Patella vulgata* (gastropoda). *J. Ultrastruct. Mol. Struct. Res.*, **102**, 147–161.

Veiss, A. (1989). Biochemical studies of vertebrate tooth mineralization. In *Biomineralization: chemical and biochemical perspectives* (ed. Mann, S., Webb, J. and Williams, R. J. P.), pp. 189–222. VCH Verlagsgesellschaft, Weinheim.

Vincent, J. F. V. (1990). *Structural biomaterials*. Princeton University Press, Princeton, NJ.

Vogel, J. J. and Smith, W. N. (1976). Calcification of membranes isolated from *Bacterionema matruchotii*. *J. Dent. Res.*, **55**, 1080–1083.

Wainwright, S. A., Biggs, W. D., Currey, J. D. and Gosline, J. M. (1976). *Mechanical design in organisms.* Princeton University Press, Princeton, NJ.

Weiner, S. (1986). Organization of extracellularly mineralized tissues: a comparative study of biological crystal growth. *CRC Crit. Rev. Biochem.*, **20**, 365–408.

Weiner, S. and Traub, W. (1980). X-ray diffraction study of the insoluble organic matrix of mollusk shells. *FEBS Lett.*, **111**, 311–316.

Weiner, S. and Traub, W. (1984). Macromolecules in mollusc shells and their functions in biomineralization. *Philos. Trans. R. Soc. London B*, **304**, 425–434.

Weiner, S. and Traub, W. (1986). Organization of hydroxyapatite crystals within collagen fibrils. *FEBS Lett.*, **206**, 262–266.

7 Morphogenesis

Morphogenesis is a general term that refers to the origin and development of *form* in biological systems. Remarkably, there seems little to distinguish between Nature's ability to sculpt structures out of soft organic materials—hearts, lungs, etc.—and hard inorganic minerals, such as bones and teeth. Indeed, as we described in Chapter 2, the propensity to produce elaborate inorganic skeletons and shells is part and parcel of the development of form in microscopic creatures, which suggests that the processes of morphogenesis and biomineralization have been strongly coupled throughout evolution.

At first sight, however, there appears to be a fundamental inconsistency between the geometric shapes adopted by inorganic crystals and the curved and twisted forms of biological minerals. The former are constrained in shape due to the symmetry of the unit cell which imposes limitations on the faces that can be expressed in a periodic structure (see Chapter 4, Section 4.8). Although some variations in the geometric morphology can be induced by additives these are unlikely to produce structures analogous, for example, to the exquisite trumpet-shaped coccolith crystals shown in Fig. 2.3 of Chapter 2. Amorphous minerals, on the other hand, lack molecular periodicity so it seems reasonable that they might adopt unusual shapes and forms. However, in the test-tube, materials such as silica precipitate as spherical particles because of the isotropic nature of the disordered structure. Thus, like their crystalline counterparts, amorphous minerals require *external processing* if the intrinsic nature of the structure is to be marshalled into a complex 3-D shape. In biomineralization, this occurs by coupling the deposition of the inorganic phase to higher levels of organization in the organism so that morphogenic force fields dominate the growth process. The biological programming is so strong that the processing is independent of the structural nature of the minerals so that both crystalline and amorphous minerals are sculpted into a wide variety of unusual life-like forms.

In this chapter, we discuss some general ideas that together provide a conceptual framework for understanding the morphogenesis of biominerals. We focus predominantly on the patterning of biominerals within membrane-bounded vesicles. It is important to stress that we are dealing with a working model with many gaps in our knowledge awaiting further research. We begin with a discussion of symmetry breaking in shape-directed growth, which then leads us to the concept of vectorial regulation and its application in pattern formation.

The coupling of mineral growth and biological organization results in the morphogenesis of life-like inorganic structures with complex form.

7.1 Symmetry breaking

A key principle in morphogenesis is that the intrinsic symmetry of inorganic structures is superseded by external force fields. This gives rise to *symmetry*

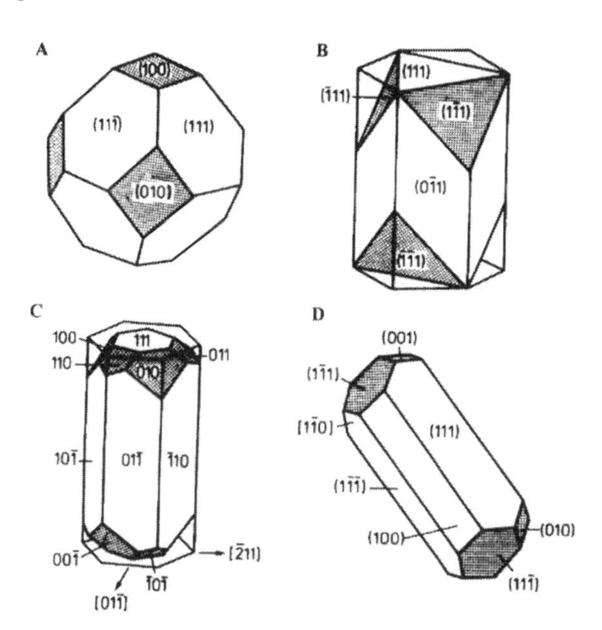

Fig. 7.1 Morphological types of bacterial magnetite single crystals. (A) cubo-octahedron, (B) and (C) hexagonal prisms, and (D) elongated cubo-octahedral.

breaking at the level of morphology such that the external shape bears no direct relationship to the structure described at the molecular scale. For example, several types of *magnetotactic bacteria* produce long elongated single crystals of magnetite (Fe_3O_4) (see Fig. 3.3 in Chapter 3, Section 3.2). Although the crystals have individual faces that can be assigned to specific Miller indices of the cubic unit cell (Fig. 7.1), they clearly do not conform to the underlying symmetry of the crystal structure, which under non-biological conditions generates particles with isotropic shape, such as the cubo-octahedron (Fig. 7.1A). Instead, the bacterial magnetite crystals are elongated specifically along only one of several possible symmetry equivalent directions. In most cases, growth of the crystals occurs preferentially along only one of four equivalent <111> axes, even though all four directions should be the same in terms of energy and structure. This has the consequence that the high symmetry of the cubic system is reduced by the biological growth mechanism with the result that the crystal morphology does not directly correspond to the lattice symmetry of magnetite (see further reading). In some bacteria, this effect is even more pronounced, with distorted bullet-shaped crystals being reproducibly produced (Fig. 7.2). Again, these crystals clearly show some faces that can be assigned to the cubic symmetry, but are further distinguished by their absence of a centre of inversion.

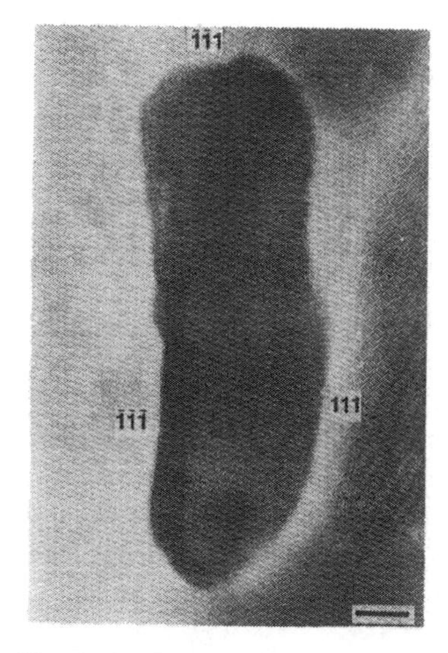

Fig. 7.2 Bullet shaped bacterial magnetite single crystal with side and end {111} faces. Scale bar, 10 nm.

It seems reasonable to assume that, all things been equal, an organism would shape a crystal in sympathy with the underlying structure in order to take advantage of naturally occurring directions of growth. For example, the intrinsic crystallographic anisotropy of calcite ($CaCO_3$) can be exploited in the growth of long spines by aligning the direction of fast crystal growth—the *c*-axis—with the long axis of the morphological form. This makes sense as the level of symmetry breaking is low and the process is presumably efficient.

But there are many other examples of calcitic sponge spicules in which the correspondence between crystallography and morphology is less predictable (see further reading). Similarly, crystalline biominerals with complex 3-D architectures, such as the curved skeleton of an adult sea urchin, often show no apparent correlation between the external morphology and crystal structure. In such cases, the crystal symmetry appears to be subsumed within a 'morphological plasticity' that is imposed by external fields arising from higher levels of organization in the organism.

In general, the breaking of symmetry at the level of morphology is *time-dependent*. For example, the elongated bacterial magnetite crystals shown in Fig. 7.2 develop initially in the shape of isotropic cubo-octahedral nanoparticles. There is therefore no symmetry breaking in the crystal morphology at the early stage of growth. Indeed, there appears to be no spatial or chemical constraint imposed on crystal growth until the crystals reach about 20 nm in size. Only then do the crystals begin to develop the elongated form. Similarly, the calcium carbonate crystals of immature *coccoliths* are rhombohedral whereas the mature structures are sculpted into complex non-regular shapes (see Chapter 8, Section 8.4, for further details). Likewise, the initial form of the Mg-calcite crystals in larval *sea urchin spicules* is based on a spindle-shaped morphology with rhombohedral end faces. In each case, the subsequent development of the morphological forms is coupled to a greater or lesser extent with specific aspects of the crystal symmetry but there seem to be no general selection rules. For example, whereas each rhombohedral coccolith crystal predominantly grows along the crystallographic *c* axis to produce a thin plate-shaped crystal that subsequently develops a hammer-headed extension (Fig. 7.3A), the spindle-shaped crystals of the larval sea urchin are restricted in their growth specifically along the *c* direction! Instead, they develop preferentially as needle-like spines along three different crystal-

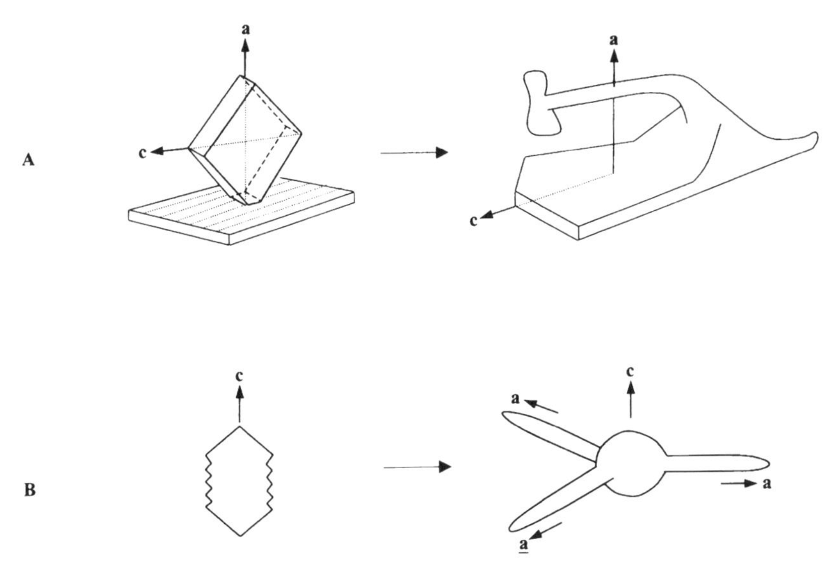

Fig. 7.3 Time-dependent changes in biomineral morphology: (A) coccolith crystal; (B) larval sea urchin spicule.

lographic *a* axes to produce a tri-radiate spicule (Fig. 7.3B; see also Fig. 4.28 in Chapter 4).

Symmetry breaking at the morphological level often occurs by time-dependent morphogenesis. The resulting mineral forms may contain relicts of the underlying crystal symmetry or show no correspondence.

7.2 Vectorial regulation

The breaking of symmetry in the morphological development of biominerals involves the imposition of biological directionality on the chemistry of the growth process. In Chapter 4, Section 4.6, we discussed how the crystal growth of inorganic minerals occurs by addition of ions and clusters to active sites, such as steps and kinks on the crystal surface, and terminates when the supersaturation level falls to the equilibrium solubility limit. In morphogenesis, these processes have to be under a higher level of control, referred to as *vectorial regulation*, in which both the extent and direction of growth vary with time according to some predetermined programme. As a general principle, we can consider the vectorial regulation of crystal growth to arise from *chemical and physical patterning* of the reaction field associated with mineralization. These patterning processes are fundamentally dependent on the secretion of shaped biological compartments, such as vesicles and organic macromolecular frameworks.

The vectorial regulation of biomineralization involves the chemical and physical patterning of inorganic deposition.

7.2.1 Chemical patterning

Chemical patterning refers to a vectorial process in which the levels of supersaturation in a vesicle or organic matrix are regulated in time and space by changes in the activity or positioning, or both, of ion pumps, channels and cells responsible for the supply of ions to the mineralization front. For example, several types of ion channels were discussed in Chapter 5, Section 5.3, and by turning these on or off in a controlled sequence that is spatially orchestrated within a shaped compartment, the supersaturation will rise and fall accordingly. Moreover, if ions flow into the localized compartment only at specific sites, then these will be the initial regions of mineral growth. If these sites are now turned off and other pumps further along the membrane are switched on, then the mineral develops along preferred directions due to the vectorial flow of the ion stream (Fig. 7.4). Under such conditions, the mineral morphology is contingent on the time and spatial dependence of the membrane transport processes that occur within the shaped environment.

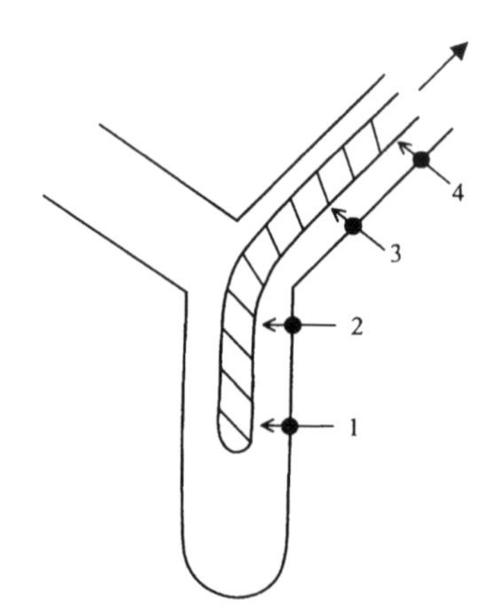

Fig. 7.4 Chemical patterning of biomineral form in a shaped compartment containing membrane pumps located at specific sites. The direction of growth is given by the time sequence, 1 to 4.

Another possibility is that the flow of ions and molecules to the mineralization front is spatially directed by the trafficking of storage granules enriched in ions such as Ca^{2+}, or kinetically stabilized as mineralized precursors such as amorphous calcium carbonate or calcium phosphate. For example, the early stages of coccolith calcification (Chapter 5, Section 5.4.2) involve the migration and accumulation of Ca^{2+}-polysaccharide-rich nanoparticles (*coccolithosomes*), initially along the lateral edges and then later on the top face of the coccolith-containing vesicle. As the nanoparticles supply calcium ions to the vesicles, these changes can influence the patterning of crystal growth inside the membrane-bounded compartment.

The coordinated flow of ions across shaped membranes is a potential mechanism in the formation of biominerals with complex morphologies.

7.2.2 Physical patterning

The physical shaping of biological compartments is of fundamental importance in the vectorial regulation of biomineral growth. In general, the mineral adopts the shape of the associated organic architecture, which often remains unchanged during the replication process, analogous to a cast produced in a mould. Alternatively, the process is dynamic with the vesicle or framework changing shape with time and the crystal morphology following on behind. This is what happens during the growth of the elongated bacterial magnetite crystals discussed in Section 7.1. Each crystal is enclosed by an organic membrane that acts as a spatial constraint on crystal growth but the nature of the restriction changes with time. The working hypothesis is that the vesicle membrane begins to elongate in a direction parallel to the adjacent cell wall once the crystals attain a size of about 20 nm. After this, the crystal growth follows and fills in the newly available space, which allows the magnetite crystals to increase in length whilst the width remains unchanged. As shown in Fig. 7.2, one end of the elongated crystal is faceted, because it originates from the original well-defined cubo-octahedral form, whereas the other is tapered and curved from direct contact with the expanding vesicle membrane. The close correspondence between the shape of the vesicle and crystal morphology also explains the curvature often seen in the magnetite nanoparticles.

Although the changes in the shape of the vesicle associated with bacterial magnetite formation are relatively straightforward—a simple elongation along one axis—in many other systems the process becomes highly complicated and gives rise to the physical patterning of complex 3-D mineralized structures. However, in each case the vesicles have to be held in place and repositioned with time in a coordinated programme of pattern formation. How this occurs is described in the following section.

Shaped biological compartments are used to physically pattern complex morphologies in biomineralization. The process may involve direct casting or emerge from dynamic changes in shape during mineralization.

7.3 Pattern formation in biomineralization

Metaphorically, the process of pattern formation in biomineralization resembles a 'chemical medusa' that transforms soft organized matter into hard stone-like structures. Figure 7.5 illustrates how the process originates from biosynthetic pathways involved with the assembly of patterned and shaped organic architectures, which when primed with appropriate macromolecules

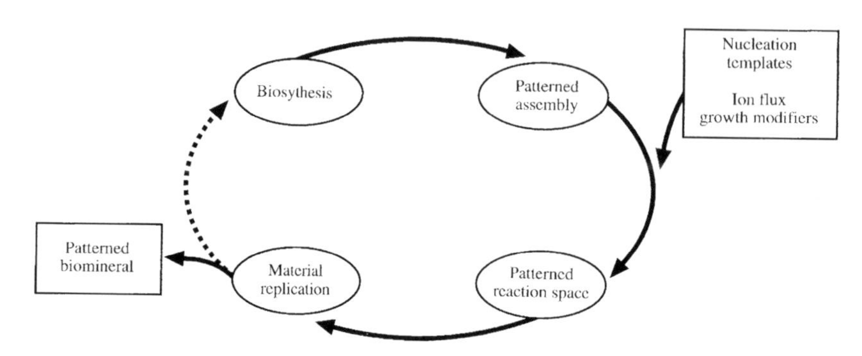

Fig. 7.5 General model of pattern formation in biomineralization.

and nucleation templates become active in biomineralization. The direct replication of the organic shape produces the fully formed biomineral in a single step. Alternatively, as shown by the dashed line in Fig. 7.5, the presence of the mineral within the reaction space generates feedback in the biosynthetic process so that changes in morphogenesis occur. For example, as a mineral grows inside a vesicle it begins to dominate the replicated organic structure, so that the shape rigidifies and there is no longer any requirement for the vesicle to be held in place. The patterning process then becomes modified and more dependent on the crystal force field than the biosynthesis of supporting structures around the vesicle. Similarly, surface interactions between the mineral and the surrounding vesicle membrane could result in the anchoring of phospholipid molecules with the consequence that the membrane loses its fluidity in certain regions. This influences the growth of the vesicle, which can now only take place by addition of new molecules to those areas of the vesicle containing unattached phospholipids. In this way, changes in the patterning process can occur through synergistic coupling of the assembled inorganic and organic components.

The general features of pattern formation in *intracellular* biomineralization are illustrated in Fig. 7.6. The dynamic shaping of a vesicle (V) involved in biomineralization (B) takes place by anchoring the lipid membrane to an underlying *scaffold* such as the cell wall, surface of an internal organelle or intracellular filament. This is achieved through the use of directing agents, usually in the

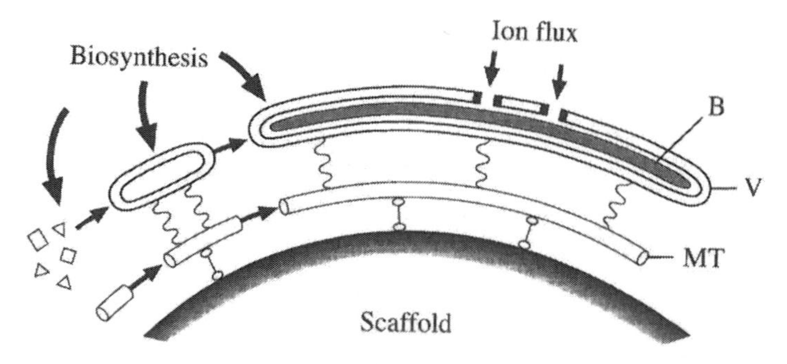

Fig. 7.6 Pattern formation in intracellular biomineralization. B, biomineral; V, vesicle; MT, microtubule.

Table 7.1 Patterning of biominerals within membrane-bounded vesicles

Scaffold	Directing agents	Form	Architecture	Example
Internal cell wall	Microtubules	Curved SiO_2 rods	Woven mesh	Choanoflagellates
	Unknown	Shaped Fe_3O_4 crystals	Linear array	Magnetotactic bacteria
	Microtubules	Curved SiO_2 shell	Hollow cyst	Chrysophytes
	Microtubules/ vesicles	Perforated SiO_2	Hollow shells	Diatoms/ radiolarians
Endoplasmic reticulum	Microtubules	Curved SiO_2 sheets	Interlinked scales	Chrysophytes
Nuclear envelope	Microtubules?	Shaped $CaCO_3$ crystals	Scales/ coccosphere	Coccolithophores
Cytoplasmic sheath	Microtubules	Reticulated SiO_2/$SrSO_4$	Micro-skeleton	Radiolarians/ acantharians
Cellular groupings	Vesicles/ microtubules	Curved/shaped $CaCO_3$	Spicules/hollow shell	Echinoderms
Cellular organization (no vesicles)	Biopolymers/ force fields	Multilevel structures (bone/shell)	Macro-skeleton	Vertebrates/ molluscs

form of *microtubules* (MT), which are tough protein cables that act as guy-ropes for the vesicle. These link the vesicle to specific sites on the scaffold so that the vesicle is stretched out like a sausage or a flattened balloon. The patterning process is extended in time by the progressive biosynthesis of the organic components and their assembly along specific directions of the scaffold. Moreover, the physical shaping of the vesicle and its associated biomineral takes place in combination with chemical patterning arising from the time-dependent vectorial flux of ions across the vesicle membrane.

Several examples of how this two-component model can be applied to various biomineral morphologies are listed in Table 7.1, and described below (see also, further reading). As a general rule, intracellular space is criss-crossed with micro-skeletal networks and associated stress fields, so the equilibrium spherical shape of a vesicle membrane can be readily distorted by mechanical and structural forces operating locally and at a distance. Empirically, it appears that the shaping of a vesicle through the coupling of microtubules and scaffolds produces two fundamental types of stress fields that act either *tangentially* along the surface of the cell wall or an internal organelle or *radially* across the cell. In many systems, the microtubules can attach to structural polymers, such as *spectrin* and *tubulin* that are used as the tangential and radial patterning agents, respectively (Fig. 7.7). Tangential patterning tends to produce biominerals with hollow-shell morphologies, while the radial force field generates mineralized spines arranged like spokes on a wheel (Fig. 7.8A). In combination, the two systems produce structures with both radial and tangential organization as illustrated by the simple but elegant example shown in Fig. 7.8B. In this structure, the mineral (silica) is patterned as radiating spines until the vesicle comes into contact with the inner surface of the cell, after which it is directed tangentially along one specific direction.

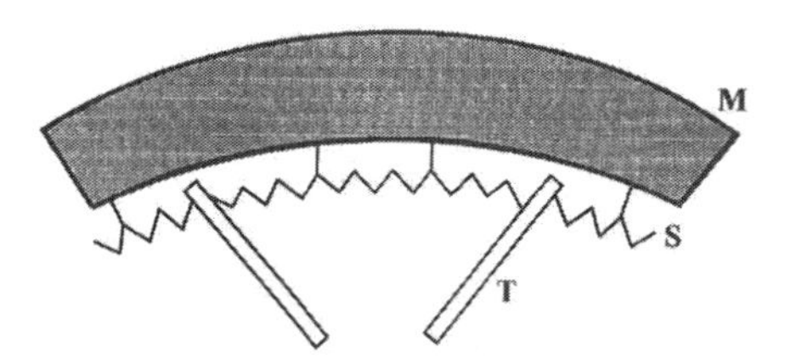

Fig. 7.7 Locations of radial (tubulin, T) and tangential (spectrin, S) structural proteins on the cell membrane (M).

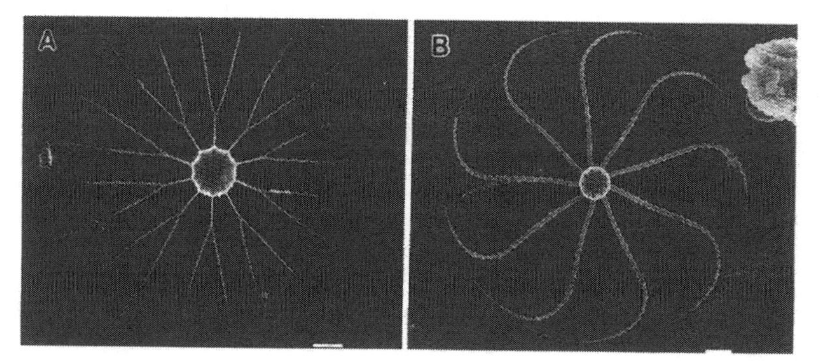

Fig. 7.8 Biomineral forms based on (A) radial and (B) radial and tangential patterning. Scale bars, 10 μm.

A similar retrospective analysis can often be used to explain the formation of more complex biomineral architectures, although the successive layering of various combinations of radial and tangential patterning tends to make it difficult to define the actual sequence of morphogenic events.

Pattern formation in biomineralization involves the radial and tangential shaping of vesicles through the use of microtubules and scaffolds.

7.3.1 Scaffolds

Organic scaffolds in biomineralization are associated either with the surface of a microscopic object within the cell or the inner surface of the cell wall membrane. In the former, spheroidal surfaces of *intracellular organelles* are recruited as structural supports for the formation of mineralized scales. For example, flattened vesicles are used to produce curved silica scales in *chrysophytes* such as the freshwater alga, *Synura petersenii* (Fig. 7.9). As shown in Fig. 7.10, each vesicle (V) is organized into a disc-like shape by short structural filaments (F) that pin one side of the vesicle to an intracellular structure called the *endoplasmic reticulum* (ER) that is otherwise engaged in sorting proteins (see Chapter 5, Section 5.1.1). This arrangement guides the development of the plate-like structure across the scaffold surface except at the rim where microtubules (M) are required to bend the edges of the disc away from the scaffold surface.

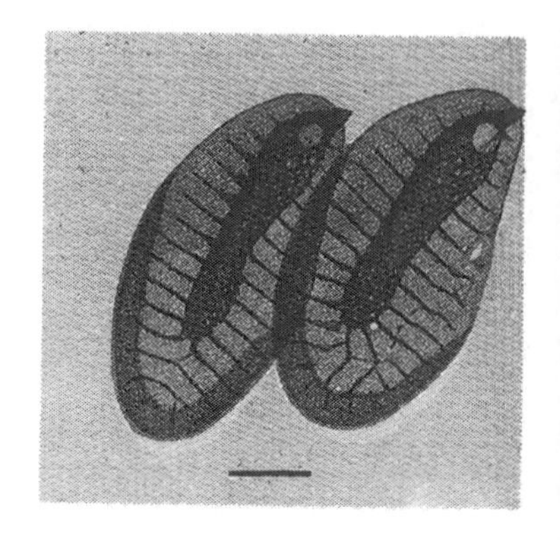

Fig. 7.9 Algal silica scales. Scale bar, 1 μm.

Similar processes are responsible for the elaborate shape of the calcium carbonate *coccolith* scales of the alga *Emiliania huxleyi* (Fig. 7.11). In this case, the coccolith vesicle (CV), which originates from the Golgi complex

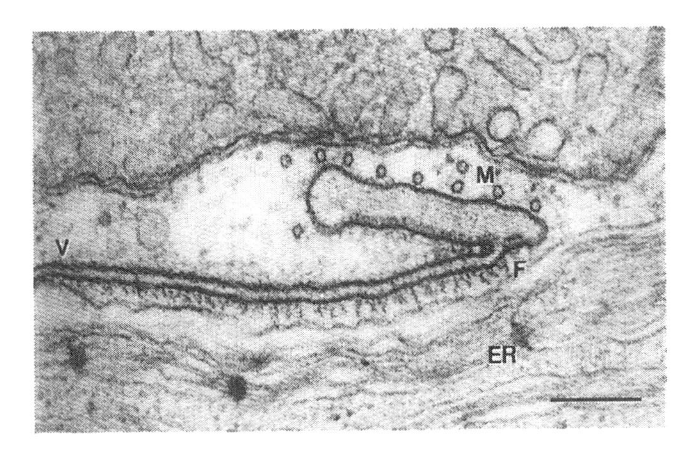

Fig. 7.10 Sectioned algal cell showing shaped vesicle (V) anchored to endoplasmic reticulum (ER) by organic filaments (F) prior to the deposition of a curved silica scale. Microtubules (M), viewed in cross-section, are also visible. Scale bar, 200 nm.

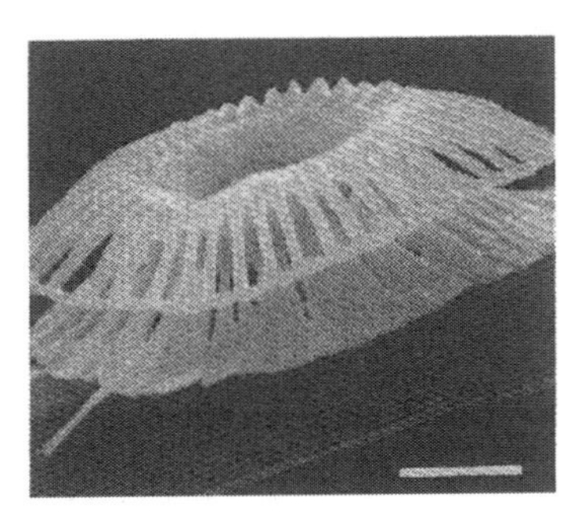

Fig. 7.11 Coccolith scale viewed approximately side-on. The scale consists of a ring of about 30 hammer-headed calcite crystals (see Fig. 7.3A) that together produce a double-rimmed structure. Scale bar, 1 μm.

(G) (see Section 5.1.1), is anchored to the surface of the *nuclear envelope* (N), as shown in Fig. 7.12. The coccolith scales (C) are sculpted and assembled inside the shaped vesicle, which then detaches from the nuclear envelope and migrates to the cell surface where it is integrated into a larger structure called the *coccosphere* (CS). Although details of the constructional stages of mineralization have been determined—see Chapter 8, Section 8.4.1, for details—the actual process by which the vesicle is shaped is currently unknown. Interestingly, no microtubules have been specifically identified in association with the coccolith vesicle or nuclear envelope.

The internal surface of *cell walls* and their associated structural polymers, such as spectrin, are often used as organic scaffolds for the patterning of vesicles involved with biomineralization. For example, as shown in Fig. 7.13, certain protozoa, called *choanoflagellates*, produce curved rods of amorphous silica (S) within elongated sausage-shaped vesicles (V) that are stretched against the cell membrane (CM) by microtubule filaments. The individual vesicles are pulled out to such an extent that each associated mineral rod has

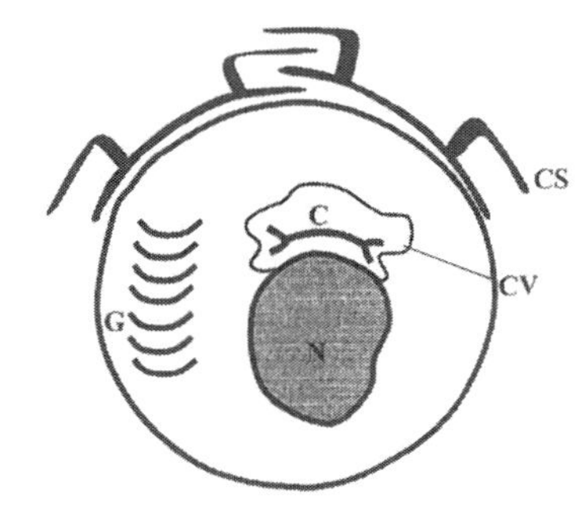

Fig. 7.12 Location of coccolith scale (C) and coccolith vesicle (CV) in algal cell. The vesicle is anchored to the surface of the nucleus (N) near to the Golgi complex (G). Part of the mature coccosphere (CS) is also shown.

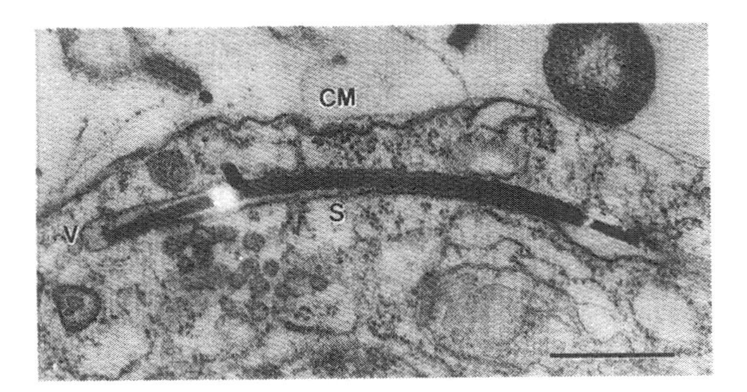

Fig. 7.13 Cell section showing curved rod of silica (S) inside a vesicle (V) shaped against the cell membrane (CM). Scale bar, 500 nm.

a radius of curvature determined by the shape of the adjacent cell membrane. When molecules such as *colchicine*, which is a known inhibitor of microtubule synthesis, are added to a growing culture of these organisms, the mineralized rods that subsequently form are bent and misshaped because the connections between vesicle, microtubules and cell wall are removed.

Many unicellular organisms use the surface of cell walls and intracellular organelles as scaffolds for the shaping of vesicles and their associated biomineralized products.

7.3.2 Vesicle foams—diatoms and radiolarians

If a vesicle spans across the entirety of the cell wall then the resulting biomineral is in the form of a closed hollow shell. This happens, for example, when algal cells enter a dormant stage in their life cycle—a long sleep—and need to protect themselves from the external environment. A cyst in the form of a thin continuous shell of amorphous silica houses the organism and is subsequently removed when the cell reactivates. Alternatively, a porous spherical shell is produced if a raft of closely packed vesicles is assembled against the cell wall. For example, the lace-like silica shells of *diatoms* and *radiolarians* are derived from close-packed arrays of large vesicles, referred to as *areolar vesicles*, that are secreted and attached to the membrane wall of the cell prior to mineralization. The vesicles are arranged into a thin foam-like film with regular polygonal symmetry and organized interstitial spaces. Significantly, mineralization occurs around but not inside the vesicles so that the void spaces are used to pattern a continuous porous framework of silica across the cell wall (Fig. 7.14).

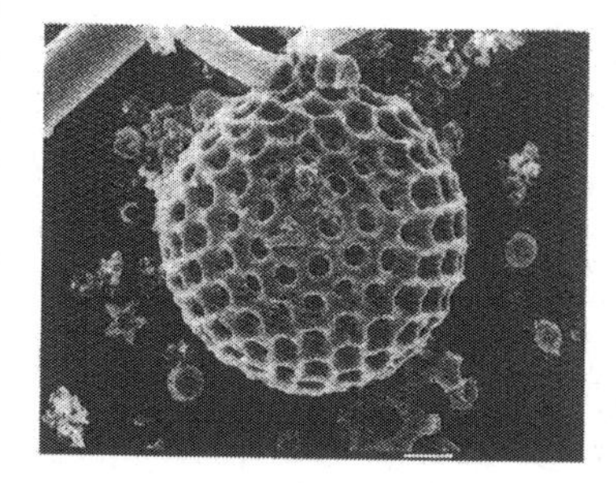

Fig. 7.14 Radiolarian silica micro-skeleton. Scale bar, 10 μm.

Using this simple model, the diversity of patterns observed in diatom shells (*frustules*) can be explained at least at a superficial level by geometrical deviations in the close packing of the vesicles against the curved surface of the cell wall. However, in reality the process is more complicated because the organization of the vesicles is not controlled primarily by surface tension as implicit in the packing model, but by specific morphogenic processes, as shown in Fig. 7.15 for silica deposition in the diatom *Coscinodiscus wailesii*. Long tubular vesicles, referred to as silica-deposition vesicles (SDV), are secreted and assembled along with microtubules against the endoplasmic

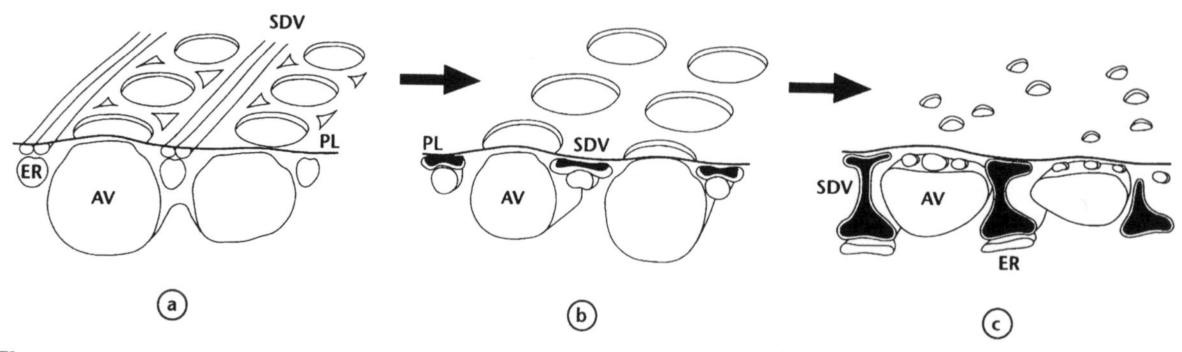

Fig. 7.15 Stages in the morphogenesis of the diatom frustule. See text for details.

reticulum (ER), which is sited in the interstitial spaces between the large areolar vesicles (AV) (Fig. 7.15A). Silica deposition is initially confined tangentially to the tubular system such that an open geometric mesh of pores, often several micrometres across, is established across the cell wall, or plasmalemma (PL) (Fig. 7.15B). In order to build the structure in 3-D, the silica-deposition vesicles extend away from the cell wall along the sides of the large areolar vesicles (Fig. 7.15C). In the final stages of biomineralization, the areolar vesicles detach and withdraw from the cell wall, and the newly formed space is infiltrated with small vesicles that are organized into elaborate patterns. Subsequent mineralization between these vesicles gives rise to a thin veil of nanoporous silica across each of the primary pores (Fig. 7.15C). Amazingly, the arrangement of the small vesicles is so precise and reproducible that species-specific nanoscale patterns are produced within each geometric cell of the silica framework (Fig. 7.16).

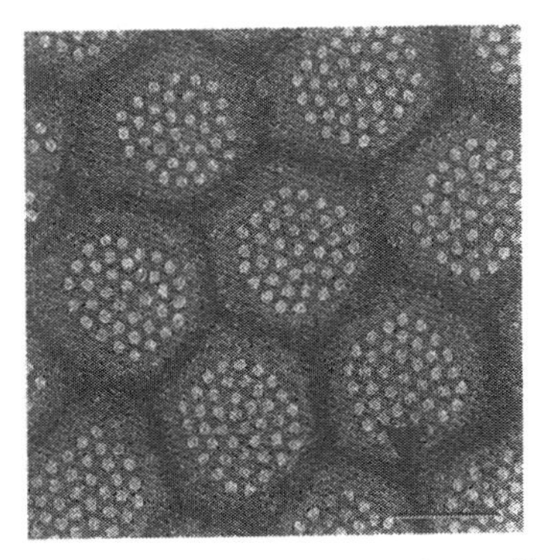

Fig. 7.16 Patterning of small holes, 40 nm across, within the hexagonal framework of a diatom frustule. Scale bar, 400 nm.

More complex intracellular architectures can be synthesized by extending the above mineralization mechanism. In particular, reiteration of the patterning process but on scaffolds of increasing diameter produces a series of concentric porous shells. These can be generated from a continuous pathway if the tangential patterning of each shell is coupled with the radial growth of structural units so that the shells are physically connected. For example, some radiolarians build multiple concentric shells of reticulated silica that are connected by numerous radial pillars of small silica spicules (Fig. 7.17). In other radiolarians, the relative proportions between tangential and radial growth are changed towards the latter so that a single porous mineralized shell with large radial spines is produced (Fig. 7.18).

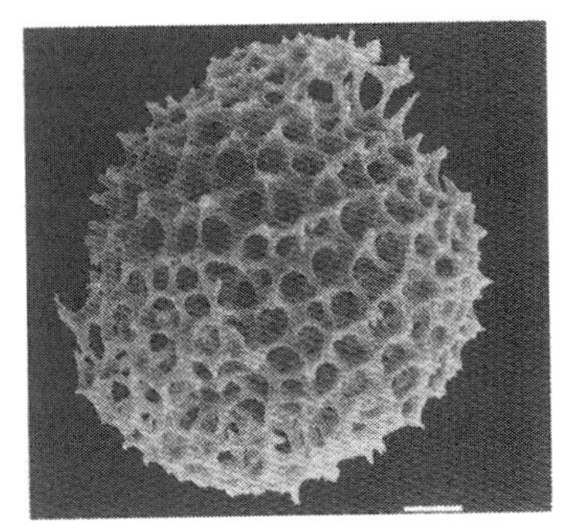

Fig. 7.17 Radiolarian micro-skeleton constructed from concentric porous shells connected by radial struts. Scale bar, 10 μm.

Similar patterning processes give rise to strontium sulfate micro-skeletons in *acantharians*. In these structures, 20 straight spicules of strontium sulfate emanate outwards from the centre of the cell (see Fig. 2.15 in Chapter 2). Each spine grows in a vesicle that is possibly attached to tubulin filaments. When the spines come into contact with the cell membrane, the vesicles and

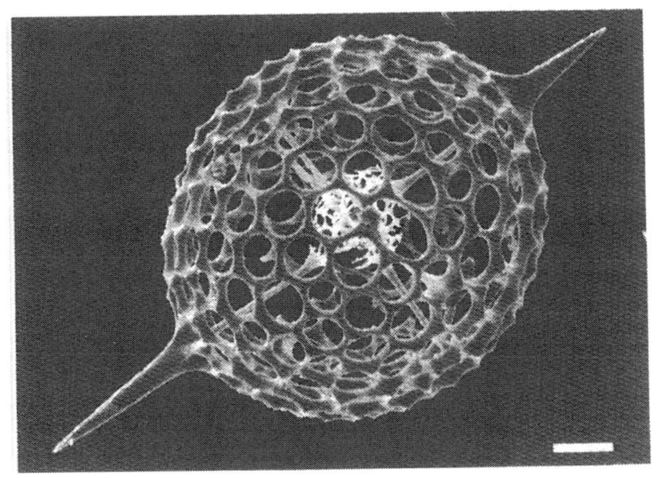

Fig. 7.18 Radialarian micro-skeleton with single porous shell and large centrally organized radial spines. Scale bar, 5 μm.

mineral begin to grow tangentially along the surface of the cell to form a thin spherical frame that interconnects with the radial struts. Unlike in the radiolarian architectures, the central spines of the acantharian skeletons are precisely organized around the central locus. And remarkably, this occurs according to a geometric rule that produces an arrangement with D_{4h} point group symmetry.

Hollow porous inorganic shells are produced in diatoms and radiolarians by patterning processes based on closely packed arrays of vesicles organized against the cell wall. The continuous silica framework results from the infilling of membrane-bounded compartments present in the interstitial spaces of the vesicle foam. Coupling of this mechanism with radial growth produces complex 3-D micro-skeletal forms in a wide range of unicellular organisms.

7.3.3 Cellular groupings

Many biominerals are patterned in simple *multicellular* organisms by processes that resemble those discussed in the previous section. Instead of foams of closely packed vesicles, groups of cells are used to pattern the micro-skeletal structures. Mineralization occurs specifically within the interstitial spaces that contain extracellular matrix filaments, frameworks, or vesicles. The simplest case involves a pair of cells, which work in concert to manufacture an extracellular linear spicule (Fig. 7.19A). Adding a third cell to the group produces a tri-radiate form (Fig. 7.19B). Both these morphologies are common in Mg-calcite *sponge spicules*. By similar reasoning, a

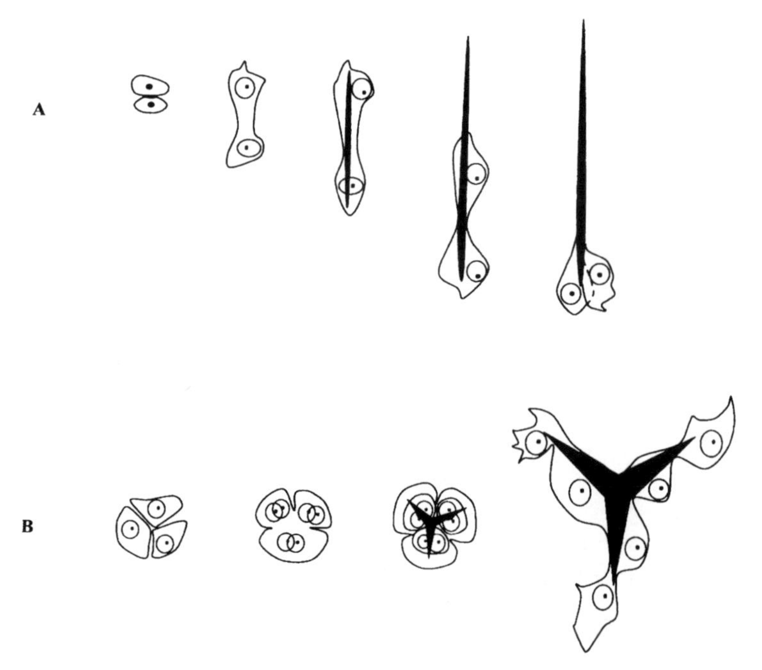

Fig. 7.19 Calcareous sponge spicules: (A) linear spine from two cells; (B) triradiate form from three cells.

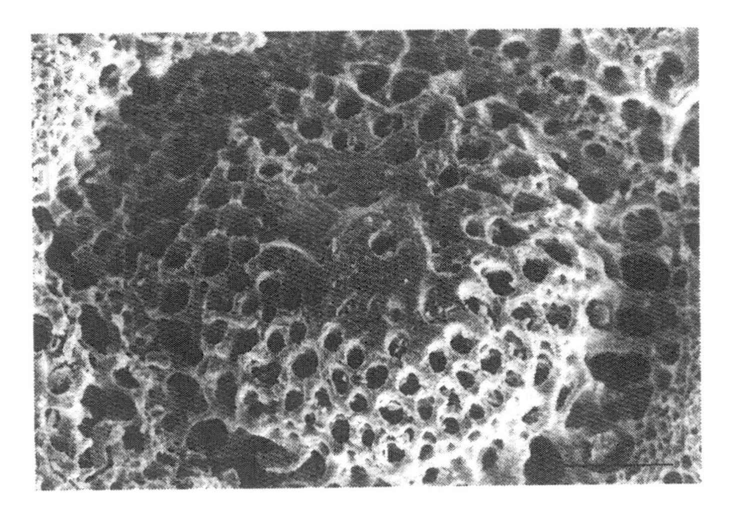

Fig. 7.20 Sea urchin skeleton. Scale bar, 50 μm.

cluster of four cells should give a quasi-tetrahedral arrangement, and this is the basis for the construction of some types of silica sponge spicules.

It follows that large-scale porous extracellular structures can arise when whole communities of cells are used collectively as a patterning template for mineralization. A striking example is the elaborate labyrinthine skeleton of the adult *sea urchin* shell that consists of an interlocked assembly of five large curved plates, each of which is in the form of a continuous porous meshwork sculpted from a single Mg-calcite crystal (Fig. 7.20). The pores are several micrometres across and are arranged in different architectures depending on the local organization of the associated cells.

Although this simple relationship between the arrangement of cells in collective structures and the shape of their associated biominerals explains at a basic level the ground rules for the adoption of particular morphologies, in reality the situation is much more complex. The model works therefore to a limited extent when applied to relatively simple multicellular organisms. But it does not account for the multilayered structures such as bone, which exhibit a number of remarkable morphological forms across a range of length scales (see Chapter 8 for details). In such cases, the activity of the cell groupings is directed by external signals, transmitted for example via hormones, rather than simple geometric packing parameters. In shell nacre, for example, this produces a laminated structure due to a patterning process based on the regular episodic deposition of matrix sheets and aragonite crystals (see Chapter 5, Section 5.1.4 and Fig. 5.17). Many variations on this theme are possible. For instance, a chamber-like architecture is produced in the aragonite shell of the cuttlefish—the *cuttlebone*—by separating the mineralized sheets with vertical pillars (Fig. 7.21). By repeating the layer–pillar patterning process hundreds of times, this produces a strong but lightweight structure that is used as a buoyancy device.

Groups of collaborating cells act as patterning templates for the production of complex biomineralized structures in multicellular organisms.

Fig. 7.21 Cuttlefish shell showing broken layer and supporting pillars. Scale bar, 100 μm.

7.4 Variations on a theme

The patterning of biomineral shape represents a compromise between the force fields of inorganic crystallization and biological organization. As we have mentioned, various combinations of these exist, ranging from morphologies that are closely related to the unit cell symmetry to those with complex non-symmetric forms. Interestingly, on close inspection it becomes clear that even between individual organisms of the same species, these complex forms are *similar* but not identical. That is, although they can be clearly recognized time and time again, they are not perfect copies. For example, most scientists with some experience in coccolith calcification would not find it difficult to distinguish between different types of coccoliths present in a sample of chalk. The hammer-headed stirrup-like morphology of the coccolith plates of *Emiliania huxleyi* (see Fig. 7.11) would, for instance, be recognizable in an optical microscope. But when viewed at higher magnification using an electron microscope, it turns out that the dimensions and morphology of each individual calcite crystal in the coccolith plate are slightly different. Moreover, each plate itself does not contain exactly the same number of crystals even though the unusual elliptical geometry is retained.

It seems likely that this morphological similarity or equivalence reflects the inherent tension in biomineral morphogenesis between predetermined genetic mechanisms and the indeterminacy of fluctuations in the surrounding chemical and physical environment. As the latter may vary with time during the lifetime of the organism, each genetically prepared biomineral has to a greater or lesser degree an additional signature written into its morphology. Clearly, in highly functional biominerals such as bones and teeth, the degree to which the structures are contingent on the external environment must be kept to a minimum. In other biominerals the dependence of function on form is much less stringent and there can be marked oscillations in the morphology, structure and composition during growth. For example, the aragonitic 'ear-stones' (*otoliths*) of fish or the branched Mg-calcite spicules of *horny corals* (Fig. 7.22) are polycrystalline composites with variable morphologies based around a common pattern. Although they form within boundary-

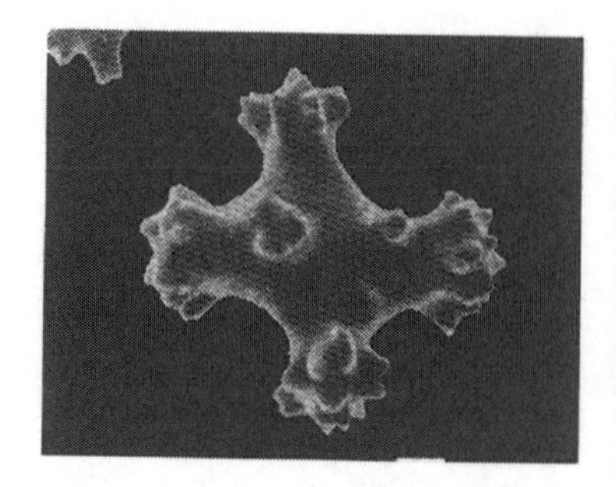

Fig. 7.22 Mg-calcite spicule from coral. Scale bar, 10 μm.

organized spaces, these are presumably subjected to local perturbations in time and space so that every spicule is slightly different.

The high fidelity of biomineral patterning takes place by programmed assembly but this is contingent on the control of indeterminate chemical and physical fluctuations in the surrounding environment. Biominerals are not identical copies but remarkably similar.

7.5 Summary

The morphogenesis of inorganic minerals with life-like forms is a distinctive feature of biomineralization. In this chapter we have described how higher levels of biological organization can impinge to such an extent on crystal growth that the underlying symmetry elements are lost from the morphological form. We discussed how this vectorial regulation can be achieved by chemical and physical patterning involving vesicles that are themselves shaped against various types of intracellular scaffolds. At the current time we have a general model based on the radial and tangential patterning of the vesicles, but the details of the mechanisms are missing in many systems.

Some progress has been made, however, on the morphogenesis of diatom frustules and radiolarian micro-skeletons. These were discussed in terms of patterning processes based on closely packed arrays of vesicles organized against the cell wall. Hollow porous inorganic shells are produced due to mineralization inside tubular compartments located in the interstitial spaces of the vesicle foam. Similar processes appear to be responsible for more complex 3-D mesh-like frames that contain centrally organized radial spines, and are also active in multicellular organisms except that groups of cells are used in place of vesicles.

At the end of the chapter we noted that the shape of many biominerals varies around a constant recognizable theme rather than being based on identical copies. Organisms work within a reasonable level of tolerance because the chemistry is susceptible to fluctuations in the surrounding environment and there need not necessarily be a strong correlation between form and function. The construction of structures, on the other hand, needs to be highly regulated if functional materials such as bone are to be used over long periods in a physically demanding environment. How these complex multilevel structures are constructed is the subject of Chapter 8.

Further reading

Aizenberg, J., Hanson, J., Koetzle, T. F., Leiserowitz, L., Weiner, S. and Addadi, L. (1995). Biologically induced reduction in symmetry: a study of crystal texture of calcitic sponge spicules. *Chem. Eur. J.*, **7**, 414–422.

Anderson, O. R. (1986), Silicification in radiolaria—deposition and ontogenetic origins of form. In *Biomineralization in lower plants and animals* (ed. Leadbeater, B. S. C. and Riding, R.), pp. 375–391. Systematics Association Vol. 30, Oxford University Press, Oxford.

Ball, P. (1999). *The self-made tapestry: pattern formation in nature*, pp. 16–49. Oxford University Press, Oxford.

Crawford, R. M. and Schmid, A.-M. M. (1986). Ultrastructure of silica deposition in diatoms. In *Biomineralization in lower plants and animals* (ed. Leadbeater, B. S. C. and Riding, R.), pp. 291–314. Systematics Association Vol. 30, Oxford University Press, Oxford.

Jones, W. C. (1970). The composition, development, form and orientation of calcareous sponge spicules. *Symp. Zool. Soc. London*, **25**, 91–123.

Leadbeater, B. S. C. (1984). Silicification of 'cell wall' of certain protistan flagellates. *Philos. Trans. R. Soc. London B*, **304**, 529–536.

Leadbeater, B. S. C. (1986). Silica deposition and lorica assembly in choanoflagellates. In *Biomineralization in lower plants and animals* (ed. Leadbeater, B. S. C. and Riding, R.), pp. 345–359. Systematics Association Vol. 30, Oxford University Press, Oxford.

Li, C.-W. and Volcani, B. E. (1984). Aspects of silicification in wall morphogenesis of diatoms. *Philos. Trans. R. Soc. London B*, **304**, 519–528.

Mann, S. (1997). Biomineralization: the form(id)able part of bioinorganic chemistry! *J. Chem. Soc., Dalton Trans.*, 3953–3961.

Mann, S. and Frankel, R. B. (1989). Magnetite biomineralization in unicellular microorganisms. In *Biomineralization: chemical and biochemical perspectives* (ed. Mann, S., Webb, J. and Williams, R. J. P.), pp. 389–426. VCH Verlagsgesellschaft, Weinheim.

Okazaki, K. (1975). Spicule formation by isolated micromeres of the sea urchin embryo. *Am. Zool.*, **166**, 567–581.

Pearse, J. S and Pearse, V. B. (1975). Growth zones in the echinoid skeleton. *Am. Zool.*, **15**, 731–753.

Perry, C. C., Wilcock, J. R. and Williams, R. J. P. (1988). A physicochemical approach to morphogenesis: the role of inorganic ions and crystals. *Experientia*, **44**, 638–650.

Preisig, H. R. (1986). Biomineralization in the chrysophyceae. In *Biomineralization in lower plants and animals* (ed. Leadbeater, B. S. C. and Riding, R.), pp. 327–344. Systematics Association Vol. 30, Oxford University Press, Oxford.

Thompson, D. W. (1942). *On growth and form*. Cambridge University Press, Cambridge.

Volcani, B. E. and Simpson, T. L. (1982). *Silicon and siliceous structures in biological systems*. Springer Verlag, Berlin.

Westbroek. P., de Jong, E. W., van der Wal, P., Borman, A. H., de Vrind, J. P. M., Kok, D. *et al.* (1984), Mechanism of calcification in the marine alga *Emiliania huxleyi*. *Philos. Trans. R. Soc. London B*, **304**, 435–444.

Wilcock, J. R., Perry, C. C., Williams, R. J. P. and Mantoura, R. F. C. (1988). Crystallographic and morphological studies of the celestite skeleton of the acantharian species *Phyllostaurus siculus*. *Proc. R. Soc. London B*, **233**, 393–405.

Williams, R. J. P. (1989). The functional forms of biominerals. In *Biomineralization: chemical and biochemical perspectives* (ed. Mann, S., Webb, J. and Williams, R. J. P.), pp. 1–34. VCH Verlagsgesellschaft, Weinheim.

8 Biomineral tectonics

The formation of higher-order structures in biomineralization involves an integrated process of construction that we refer to as *biomineral tectonics*. The word 'tectonics' originates from the Greek, *tekton*, a builder, and is often used by geologists to describe large-scale structural aspects of the earth, such as the continental plates. Chemists, on the other hand, have used the term 'molecular tectonics' to describe the rational construction of molecule-based structures from discrete well-defined building blocks. In a similar vein, this chapter considers the construction of complex multilayered biomineralized structures in terms of the integration of preformed building blocks or modules across a range of length scales. For this, we need to bring together the principles and concepts discussed in Chapters 4 to 7, so that the chemical, spatial, structural and morphological aspects of biomineralization are assimilated into a general process of biomineral tectonics (see Table 3.2 in Chapter 3, Section 3.6). We begin with a description of bone as an archetype of a higher-order structure in biomineralization.

Biomineral tectonics refers to an integrated process of construction that extends across many length scales and results in higher-order structures.

8.1 Structural hierarchy–bone

The properties of bone are a list of apparent contradictions; strong but not brittle, rigid but flexible, lightweight but solid enough to support tissues, mechanically strong but porous, stable but capable of remodelling, and so on. These conflicting demands are met by six hierarchical levels of structure in bone that together produce an embedded architecture which extends from the molecular to the macroscopic scale (Fig. 8.1). The building blocks include:

Level 1	1 nm	tropocollagen filaments ↓
Level 2	100 nm	collagen fibrils/mineral nanocrystals ↓
Level 3	1 μm	sheets of collagen fibrils ↓
Level 4	10 μm	concentric layers of collagen sheets/bone cells (the osteon) ↓
Level 5	1 mm	various osteon-based microstructures ↓
Level 6	> 1 mm	bone macrostructure

We have already described the first two of these levels in Chapter 6, Section 6.3.1, where we discussed how the self-assembly of a staggered array of triple helical tropocollagen filaments (level 1) resulted in collagen fibrils and

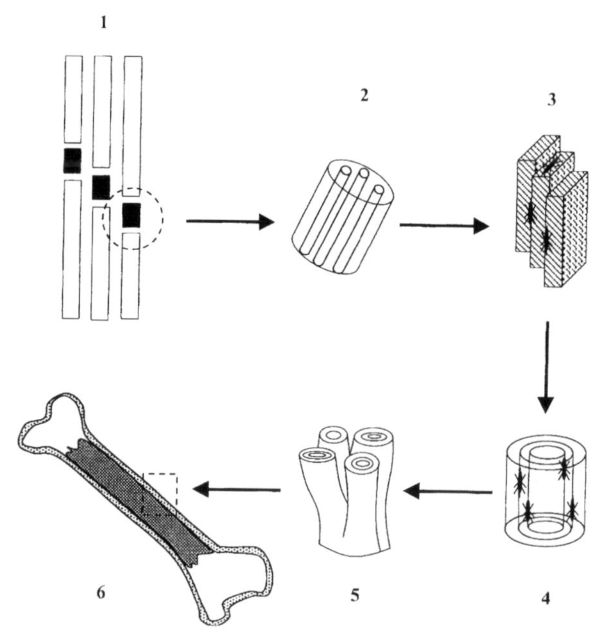

Fig. 8.1 Structural hierarchy in bone. See text for details. Circle in (1) shows the mineralization environment with hydroxyapatite crystal (black rectangle) within the hole zone of staggered tropocollagen molecules. Osteocytes are shown in (3) and (4) as black markings.

their associated hydroxyapatite crystals (level 2). As shown in Fig. 8.1, the collagen fibrils are in turn organized into sheets (level 3), although another possibility is that they are arranged into a woven texture. The sheets are assembled with bone cells (*osteocytes*) that reside between adjacent layers, and together these are either stacked in parallel arrays (*lamellar* bone) or concentrically arranged into a cylindrical structure referred to as the *osteon* (level 4). Each layer of the osteon has its constituent fibrils oriented in alternate directions like plywood. On a longer scale, the osteons are grouped together into long bundles (*haversian* bone) that are the basic building block of various bone microstructures (level 5). These include highly porous frameworks (*cancellous* bone) and more dense architectures, such as found in *cortical* bone. Each of these has a similar underlying building block but different spatial organizations according to certain structure–function relationships within the whole bone. Finally, at the macroscopic level, each bone has a specific shape and structure (level 6) so that the skeleton can work as an integrated whole.

The macroscopic shape of different types of bone is determined by *cellular differentiation* during formation of the embryo. For example, the long bone of the leg is shaped prior to mineralization in the form of a soft biodegradable model that is made from a non-mineralized highly hydrated *cartilage* matrix (see also Chapter 6, Section 6.3.2). As shown in Fig. 8.2, the cells first deposit the cartilage preformer (C), then assemble around it and differentiate into *osteoblasts* (O) that secrete collagen against the surface of the mould. Bone mineralization occurs in the collagen to give a mineralized

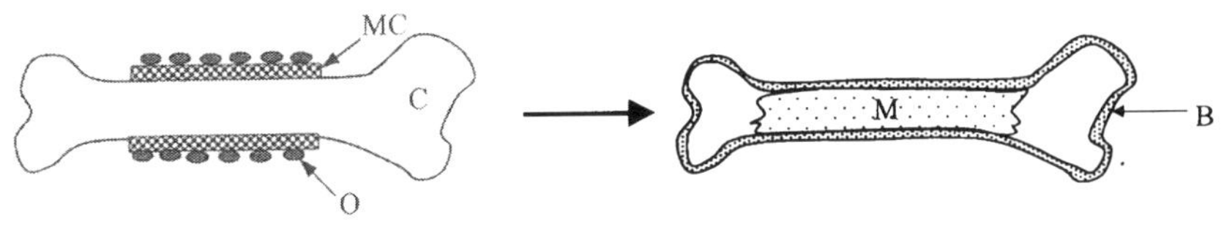

Fig. 8.2 Macroscopic shaping of long bones. C, cartilage preformer; O, osteoblasts; MC, mineralized collar; M, marrow; B, bone. See text for details.

collar (MC) around the cartilage model. Interestingly, the bone collar starves the cells inside the cartilage matrix causing degradation and the formation of a hollow cavity filled with marrow (M). The newly formed bone (B) is then remodelled into different microstructures as described above.

Bone is a hierarchical material with several levels of embedded structures that extend from the molecular to macroscopic scale. Cells and sheets of mineralized collagen represent the basic building components of various microstructures. The overall macroscopic shape is determined in the embryo by a biodegradable cartilage model.

8.2 Prefabrication

The structural hierarchy of bone is based on the ability of multicellular organisms to assimilate organic matrix-mediated mechanisms of biomineralization (Chapter 6) into long-range processes of biological organization. The use of macromolecular frameworks along with intercalated cells addresses the problem associated with building large extended structures that are many times bigger than the individual cells. Another possibility would be to use prefabricated mineral building blocks and assemble them *in situ.* For this, vesicles and their associated mineral products produced by boundary-organized biomineralization (Chapter 5) have to be integrated into cellular processes that increase the scale and complexity of construction. Moreover, because vesicles in association with other aspects of the cell—scaffolds, microtubules, etc.—can be used for sculpting discrete complex forms, such as curved spines and plates (Chapter 7), then the prefabricated mineral building blocks come in specific shapes and sizes.

In general, intravesicular mineral deposits can be moved around the cell to specific locations or ejected through the cell membrane to extracellular sites, where they are then spatially organized into higher-order structures. A fundamental concept in vesicle-based processes of biomineral tectonics therefore involves the *translocation* of prefabricated mineral-based building units. In some cases, the mineral-containing vesicles are used simply as a biological freight train, packaging and delivering their goods to an extracellular building site where they are released but not directly assembled into an ordered structure. For example, the formation of *woven* bone and calcified *turkey tendon* is associated initially with needle-shaped hydroxyapatite crystals that are encapsulated within lipid bilayer vesicles, referred to as *matrix vesicles.* The vesicles are assembled in the extracellular space some distance away from the mineralization site and become progressively filled with hydroxyapatite as

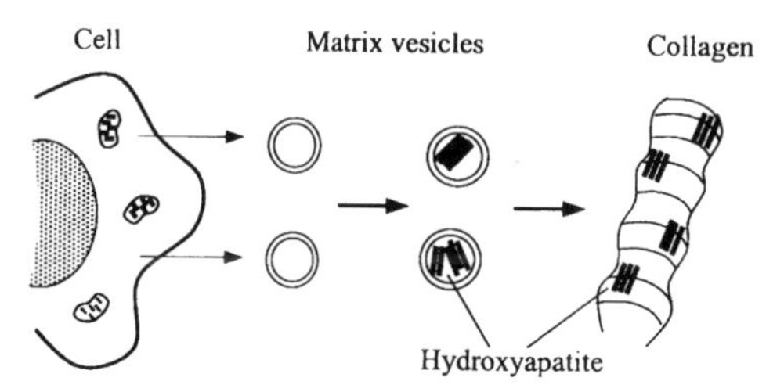

Fig. 8.3 Matrix vesicles in bone biomineralization.

they get closer to the collagen matrix of the tissue. At the mineralization front they rupture and release their contents to the collagen matrix (Fig. 8.3). What happens next is a topic of much controversy. Do the released crystals slot into the holes and grooves between the collagen fibrils? Or do they dissolve and increase the local supersaturation such that with time crystals begin to nucleate in the hole zones of the collagen fibril? The consensus appears to lie with the dissolution mechanism although scientists are still arguing passionately about the issue.

In unicellular organisms, however, there is much evidence to indicate that mineral-containing vesicles are unequivocally used as prefabricated building blocks for the construction of higher-order structures. In this process, the vesicles are involved not only with the shaping of discrete mineralized structures and their transport to a remote site but also the release of these units for incorporation into architectures that often encompass the entire cell. For example, single-celled organisms called *foraminifera* construct a mineralized shell of interconnected chambers, and in the shell of *Calcituba polymorpha* this occurs by transport and release of intravesicular Mg-calcite crystals. The needle-shaped crystals are 1 to 2 μm in length and, as shown in Fig. 8.4, packaged in bundles inside large vesicles (V) that are transported from the cytoplasm (C) to a large compartment enclosed by a thin outer organic sheath (OS). The crystals are released as intact building blocks into the mineralization front so that the shell wall (S) gradually increases in thickness. Although the first crystals to arrive are physically aligned against the surface of the organic sheath, subsequent crystals are randomly arranged within the wall.

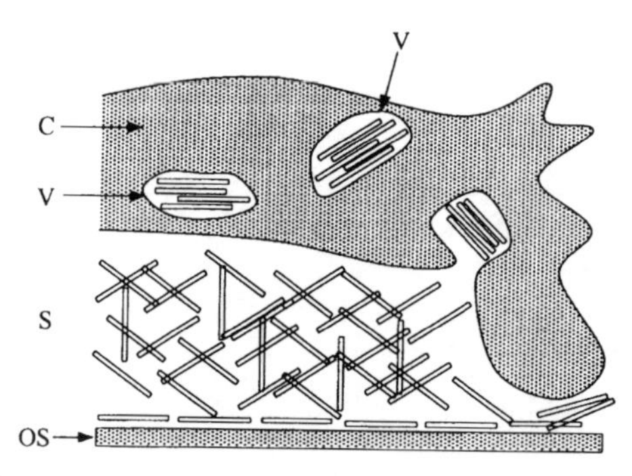

Fig. 8.4 Shell formation in *C. polymorpha*. V, vesicles containing needle-shaped Mg-calcite crystals; C, cytoplasm; OS, organic sheath; S, shell wall of Mg-calcite crystals.

Vesicles that contain prefabricated biominerals can be translocated to intracellular and extracellular sites for incorporation into larger-scale architectures.

8.3 Higher-order assembly

In many unicellular organisms, vesicles are not only used to transport mineral cargo but also deliver customized building blocks that are assembled into organized architectures. For example, although curved rods of silica are initially formed within intracellular vesicles in certain types of *choanoflagellates* (see Fig. 7.13 in Chapter 7, Section 7.3.1), they end up outside the cell as building blocks of a remarkable open-ended basket-like framework, referred to as the *lorica* (Figs 8.5 and 8.6).

How does the construction of the lorica take place? Fortunately, the cell is several micrometres in size, so high-resolution optical microscopy can be used to watch the protozoan at work. It turns out that the mature cell is firmly attached to a completed lorica and spends most of the time swimming around using the miniature trawler-net to catch bacteria. Although the lorica is intact, the cell continues to produce supernumerary rods that are shunted out into the extracellular space where they are stored towards the open end of the basket in small aggregates that look like bunches of microscopic bananas (Figs 8.7A and B). Once a full complement of rods has been accumulated—about 150 or so—then cell division occurs and, as shown in Figs 8.7C and D, the daughter cell along with the supernumerary rods squeezes through the open end of the intact lorica and out into the environment. Within three minutes, a new lorica is constructed around the daughter cell through a series of events involving the assembly and placement of the rods by extracellular tentacles (Fig. 8.7E). Although the details are not known, it appears that the basket-like framework

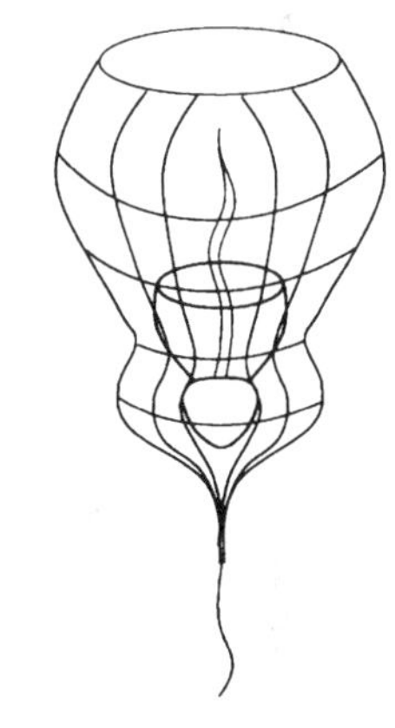

Fig. 8.6 Drawing of open-ended silica lorica with attached cell and tentacles.

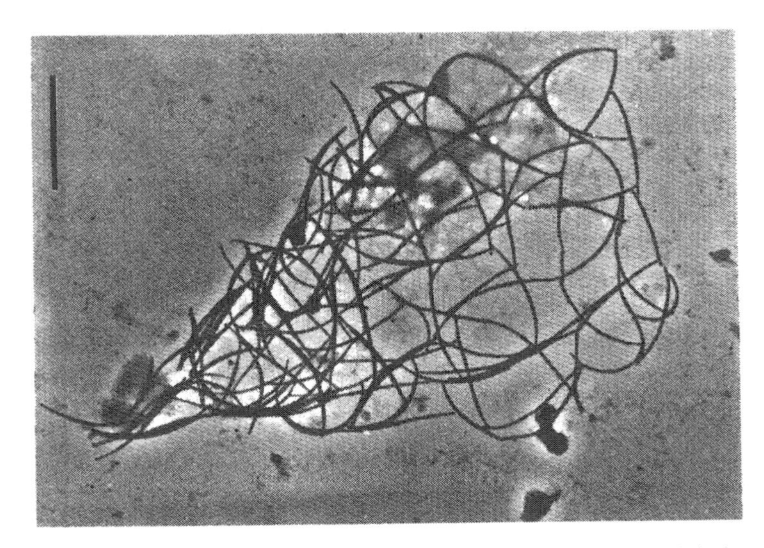

Fig. 8.5 Lorica of interconnected silica rods. The cell has been destroyed during sample preparation. Scale bar, 8 μm.

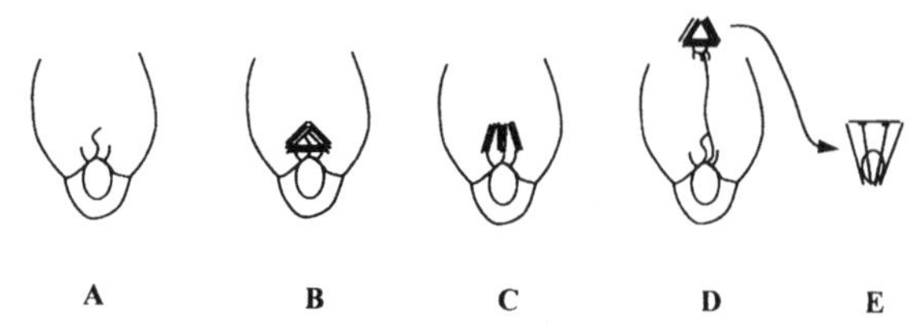

Fig. 8.7 Lorica assembly during cell division. See text for details.

is built in a series of stages that are progressively propelled away from the cell in a telescopic fashion. During this process, the individual rods slide sideways one by one out of the aggregated bunches, and become locked into place within the framework. Remarkably, the rods are 'glued' together either end-to-end or in the form of T-junctions (Fig. 8.8) by a mortar of silica and organic material of unknown composition.

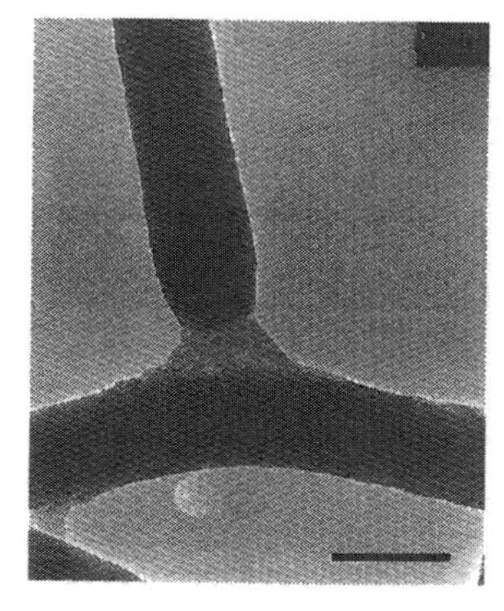

Fig. 8.8 T-join between two silica rods in an intact lorica. Scale bar, 80 nm.

Interestingly, the silica rods are metastable and slowly redissolve in the seawater to produce intact hollow rods rather than surface-etched strips. Preferential dissolution from the centre rather than the surface of the rods has a functional advantage in that it extends the time over which the higher-order structure is mechanically stable. This is because within certain boundary conditions there is only a relatively small decrease in mechanical strength between a filled and a hollow cylinder. For a similar reason, tubular rather than solid steel rods can be used in building scaffolding.

Besides mineral rods and spines, intracellular scales and plates are also used by unicellular organisms as custom-built construction units for higher-order assembly. For example, the curved silica scales of chrysophytes (see Fig. 7.9 in Chapter 7, Section 7.3.1) are organized across the cell wall like tiles on a roof. Similarly, coccolith plates are assembled into remarkable hollow-shell architectures, which we discuss in detail in the next section.

The controlled placement of vesicle-derived biominerals in organisms such as protozoa and algae produces extracellular assemblies with higher-order structure.

8.4 Multilevel processing

As a general principle we can consider the construction of biominerals with structures that transcend a single level of scale to involve a coordinated and integrated sequence of processes. To illustrate this principle of *multilevel processing*, we now discuss the various stages of construction that lead to the formation of the calcite *coccosphere* produced by the unicellular marine alga *Emiliania huxleyi*. Using this example we then develop a general framework for understanding the stages of construction associated with biomineral tectonics.

Biomineral tectonics involves a multilevel process that give rise to a sequence of construction stages integrated across various length scales.

8.4.1 Coccoliths

The calcite coccosphere produced by the unicellular marine alga *E. huxleyi* consists of a hollow shell of interconnected oval-shaped plates, called *coccoliths* (see Chapter 7, Section 7.3.1). An intact coccosphere, 6 μm in diameter, is shown in Fig. 8.9, which also illustrates how the interlocking of the double-rimmed coccolith discs stabilizes the higher-order assembly outside the cell wall. In our previous discussions in Chapter 5 (Section 5.4.2) and Chapter 7 (Section 7.3.1), we noted that the coccoliths originate from intracellular vesicles that migrate to the cell membrane where they are released and incorporated into the coccosphere. In this section, we examine the various stages of coccolith construction in detail and in so doing highlight the complexity of biomineral tectonics in unicellular organisms.

Let's start with the *E. huxleyi* coccosphere and peel away the structure to reveal the construction process in reverse. The first thing we notice is the elaborate microstructure of the 3 to 4 μm-long coccolith plates. Each scale is elliptical and consists of 30 to 40 crystalline units that together produce a tubular-shaped ring with various radial growth extensions. Viewed from outside the cell, the extensions take the form of an ordered radial array of hammer-headed elements (Figs 8.9 and 8.10B). This orientation is referred to as a distal view, and the extensions are called *distal shield elements.* If we turn over the scales so that they are cell-side up, then another set of ordered radial extensions are seen, but in this case they are plate-like and more closely spaced (Fig. 8.10A). These are termed the *proximal shield elements.* Both the proximal and distal shield elements are directed outwards from the bottom and top, respectively, of the vertical wall that comprises the *tube elements* of the central ring (Fig. 8.10; see also Fig. 7.11 for a side view). Finally, as shown most clearly in the distal view of Fig. 8.10B, there are also inwardly directing radial extensions, referred to as *central area elements*, that appear to be less ordered and fill in the central space inside the tubular ring.

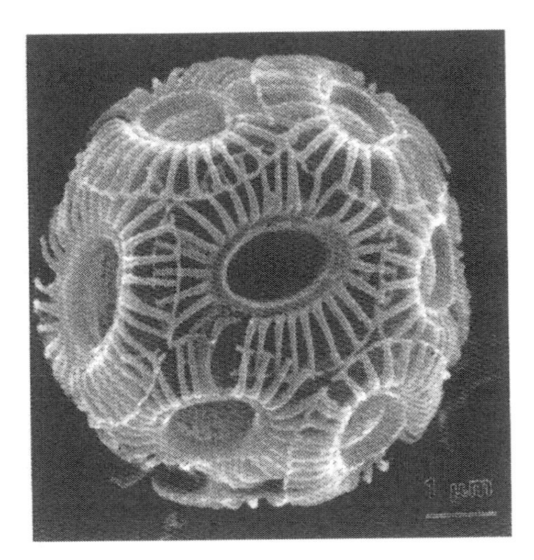

Fig. 8.9 Coccosphere of *E. huxleyi*. Scale bar, 1 μm.

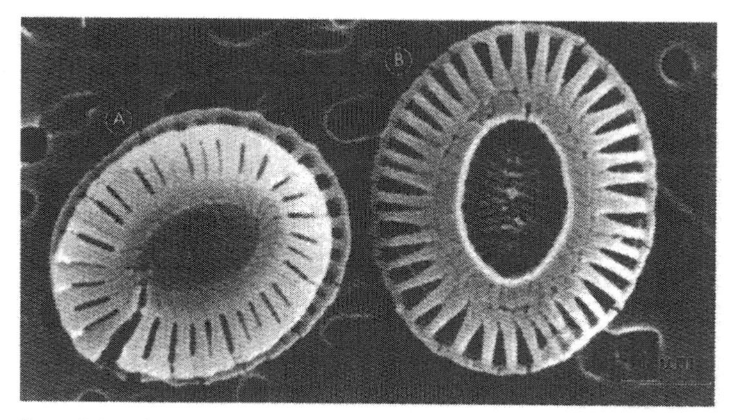

Fig. 8.10 Coccoliths of *E. huxleyi*: (A) proximal, and (B) distal view. Scale bar, 1 μm.

The next step in our analysis involves the dismantling of a single coccolith plate. For this, a dispersion of the scales in water is sonicated for a few minutes. The treatment releases individual units of the scale, which are micrometre-sized crystals with an unusual morphological form that can be related precisely to the structural elements seen in the complete coccolith (Fig. 8.11). Simply arranging the units around an elliptical ring such that the proximal shield elements lie perpendicular to the local tangent produces the double-rimmed coccolith scale, as shown in Fig. 8.12.

A key question now arises: what is the crystallographic nature of such strange microscopic objects? Because the individual units are so small, electron diffraction has to be used, and it turns out that each unit is a single crystal of calcite with the crystallographic *c* and *a* axes aligned radially and vertically, respectively, to the plane of the coccolith plate. With regard to each individual element, the *c* axis is aligned along the direction of elongation of the hammer-headed extension of the distal shield elements, and lies in the plane of the plate-like proximal shield element. Interestingly, the proximal

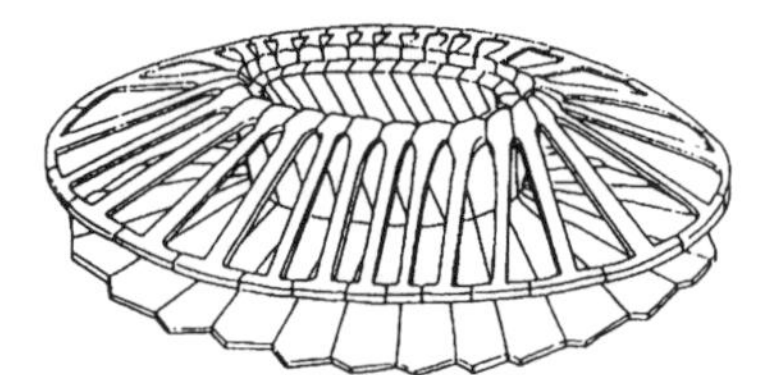

Fig. 8.12 Drawing of coccolith based on an elliptical array of discrete structural units.

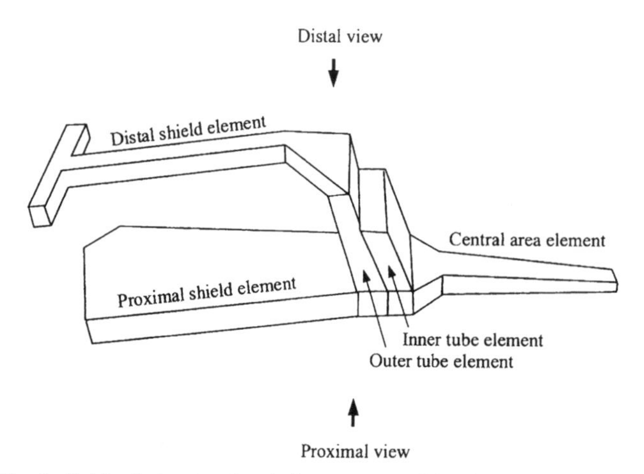

Fig. 8.11 Individual structural unit from a coccolith plate showing calcite single crystal with various growth elements.

shield element exhibits two well-defined end faces (Fig. 8.11), a large one with a Miller index of $\{\bar{1}08\}$ and a smaller one labelled as $\{104\}$, and these are bisected by the *c* axis. Because the faces are of different size, the *c* axis is not exactly perpendicular to the local tangent but offset by an angle of 20°. Amazingly, this offset always occurs in the same direction for all the units arranged around the elliptical ring so that when viewed distally the *c* axes precess in a clockwise direction with the smaller $\{104\}$ end face of the proximal shield elements always to the right-hand side. This morphological handedness makes the coccolith scale chiral, that is, mirror images of the structure are non-superimposable.

Having established the crystallography of the single-crystal building blocks of the mature coccolith scales, our final task rests with elucidating how these crystals grow and ultimately nucleate to produce such precisely oriented higher-order assemblies. Clearly, we require an adequate source of immature coccoliths, but this is not readily available because such structures are flimsy and usually hidden among all the intracellular bits and pieces. By studying coccoliths in the electron microscopy using a needle-in-a-haystack approach, very early stages of the construction process have been elucidated recently. It turns out that the assembly of the ring occurs in the very early stages of coccolith development and that a necklace of interlinked calcite crystals with geometric shape is established prior to the elaborately shaped elements (Fig. 8.13). Initially, each discrete crystalline unit is in the form of a 40-nm-thick rhombohedral plate inclined vertically to the plane of the *proto-coccolith* ring. The plate grows to a height of 100 nm and develops short radial outgrowths along the *c* axis from the top and bottom faces to produce a ring of Z-shaped crystals (Fig. 8.14A) that correspond to the cycle of tube elements in the mature coccolith. Interlinking of the Z-shaped units then occurs by selective growth along the inside rim so that neighbouring crystals are overlapped and locked against each other (Figs 8.14B and C). Because the lateral extension is always in the same direction—anticlockwise when viewed distally—the construction process now develops a morphological handedness that works through into the ultrastructure of the mature coccoliths. Further development of the proto-coccolith ring involves radial outgrowth from the

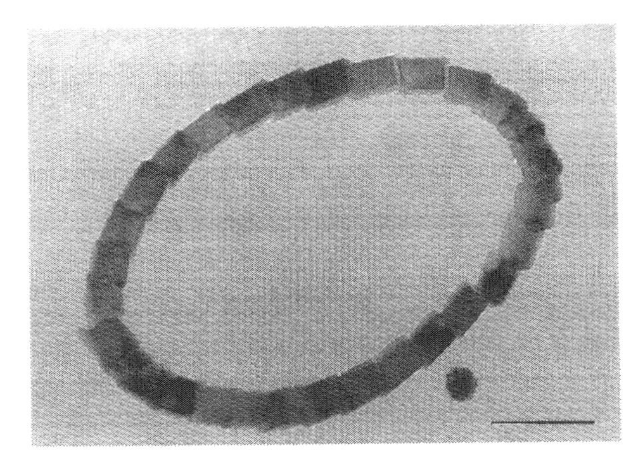

Fig. 8.13 Proto-coccolith ring of *E. huxleyi*. Scale bar, 500 nm.

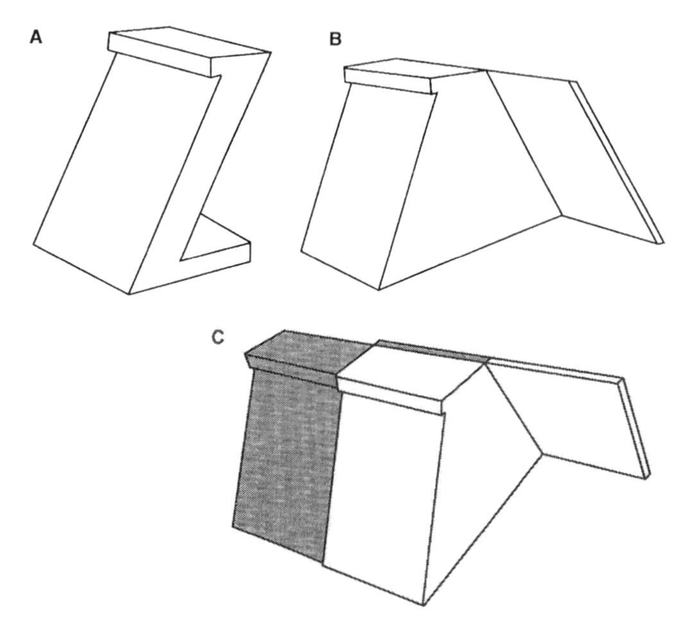

Fig. 8.14 Structural units in the proto-coccolith ring: (A) initial Z-shaped crystal; (B) lateral extension and in-filling; (C) interlocking to give cycle of tube elements.

interlocked cycle of tube elements to produce the central area, proximal and distal shield elements. The central area elements develop from the inside rim and so have a different handedness to the proximal and distal shield elements, which emanate from the base and top of the outer surface, respectively.

But we still haven't addressed the question of what happens at the very beginning—the nucleation stage. Now we are in uncharted territory with little to guide us to an answer to how the Z-shaped crystals are precisely oriented around the proto-coccolith ring such that the unit cell is always aligned with the *c* and *a* axes radial and vertical, respectively, to the plane of the ellipse. We know that each crystal is contained within a vesicle and that the vesicles are assembled around the rim of an organic base-plate. One possibility is that the rim of the base-plate contains functional groups, which induce the oriented nucleation of calcite crystals by a process of interfacial molecular recognition (see Chapter 6, Section 6.7.1). In order to account for the observed crystallographic orientations, the {110} face, which lies perpendicular to the vertically aligned *a* axis, must be preferentially nucleated against the organic base-plate and oriented such that the *c* axis, which lies in the plane of the face, is positioned radially (Fig. 8.15). Moreover, as the immature crystals are also vertically aligned with respect to the proto-coccolith ring, a potential candidate for the original nucleation surface is the small face at the base of the Z-shaped crystals shown in Fig. 8.14A.

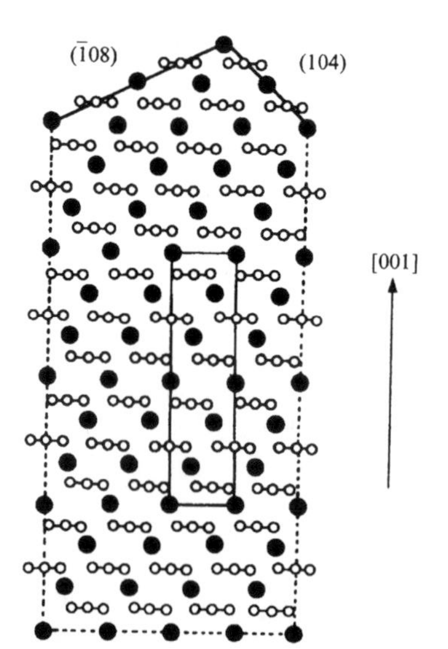

Fig. 8.15 Drawing of the molecular structure of the {110} face of calcite. Filled circles represent Ca^{2+} ions, open circles are CO_3^{2-} ions viewed side-on in projection. The shape of a proximal shield element is superimposed on the drawing, which also shows the position of the *c* ([001]) axis and the well-defined ($\bar{1}$08) and (104) end faces that lie to the left and right, respectively, in distal view.

Finally, there is one other remarkable aspect of nucleation of the coccolith structure that we need to consider. Light microscopy studies on coccoliths from other species have often revealed two distinct cycles of alternating crystals in the mature structure. One of these has all the calcite crystals aligned with their *c* axes radial (R)—as observed in *E. huxylei*—while the *c* axes are

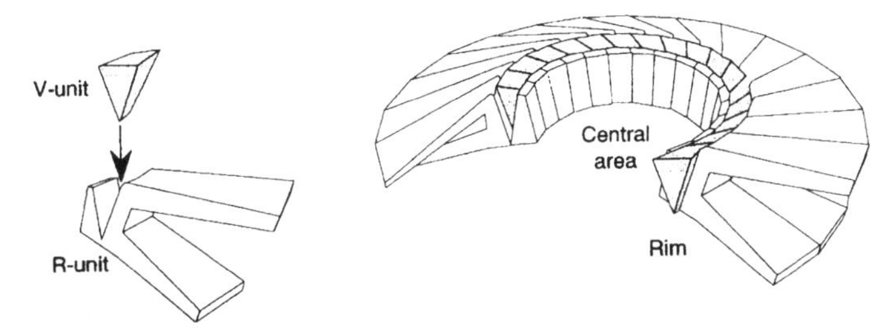

Fig. 8.16 R- and V-units in coccolith from *Watznaueria*.

vertical (V) to the plane of the coccolith ring in the second cycle. For example, mature coccoliths formed in the extinct species *Watznaueria* consist of a cycle of elaborately shaped R-units with each crystal interspaced with peg-like V-units (Fig. 8.16). However, until a few years ago nothing was known about how these observations on mature coccoliths related to the stages of construction involving the proto-coccolith ring. In this respect, *E. huxleyi* looks like a suitable system to study because it grows in culture and the coccoliths are relatively small and thin for electron microscopy imaging. But as studies on the mature coccoliths indicated only a single cycle of R-units, nobody bothered to look. Until recently that is, when by chance a few images of proto-coccolith rings of *E. huxleyi* clearly showed the presence of both R and V alternating units (Fig. 8.17). The V-units, shown by arrows in Fig. 8.17, are much smaller than the Z-shaped R-units, and grow only to a size of 60 nm in length before they become embedded in the structure of the proto-coccolith ring by preferential growth of the radially oriented crystals.

These observations indicate that the nucleation process in coccoliths is even more complex than initially thought because now we have a situation where adjacent calcite crystals in the proto-coccolith ring are nucleated orthogonal to each other. That is, within a repeat distance of 100 nm, calcite nucleation is switched back and forth through an angle of 90°! The resulting V/R alternating

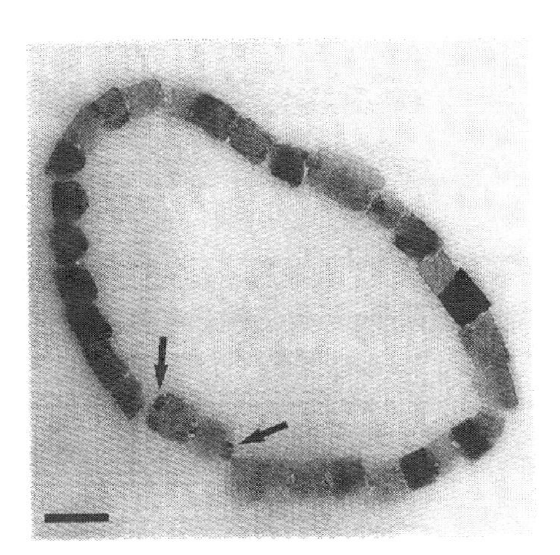

Fig. 8.17 R- and V-units in proto-coccolith ring of *E. huxleyi*. Some of the small V-units are denoted by arrows. Scale bar, 200 nm.

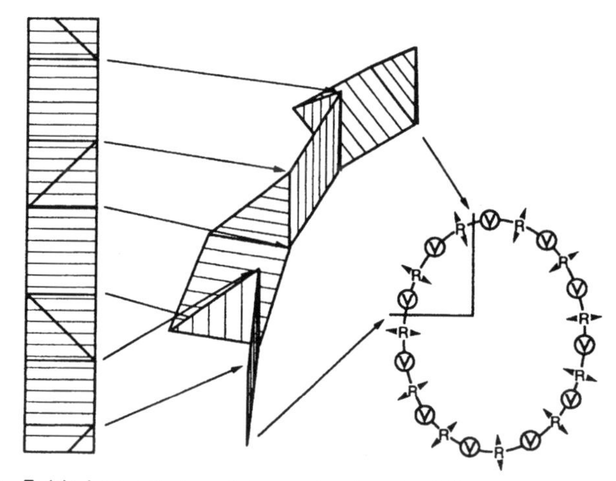

Fig. 8.18 Folded organic template mechanism for V/R nucleation in coccoliths.

pattern appears to be a general characteristic of coccolith evolution, although in *E. huxleyi* the V-units are relicts of a construction process that overtly favours development of the R-units. At the supramolecular level it is conceivable that the organic matrix responsible for nucleation of the {110} face could be folded such that the *c* axes of adjacent crystals are rotated through 90° along the same template (Fig. 8.18). This would maintain the interfacial matching between organic functional groups and ions in the {110} face of each crystal but physically change the spatial relationships between the individual units.

Coccolith construction in E. huxleyi involves a series of stages beginning with the radial (R) and vertical (V) alternation of the crystallographic c axes of calcite crystals in the proto-coccolith ring. Oriented nucleation of the {110} face on a folded organic matrix has been proposed as a mechanism to explain the V/R model. Each crystal is assembled inside a vesicle associated with the rim of an organic base-plate but only the R-units develop into morphologically complex mature units. This process involves the lateral interlinking of Z-shaped units to produce a morphological handedness and a chiral tubular ring. Radial outgrowths from the top and bottom surfaces of the tube give rise to the distal and proximal shield elements, respectively, which together produce the double-rimmed disc of the mature coccolith scale. The scales are translocated to the cell surface and interlocked into the hollow shell of the coccosphere.

8.4.2 Stages of construction

The 'total synthesis' of a coccosphere is clearly a very sophisticated multi-level process involving several constructional stages that extend across various length scales. The algal cell is not specialized for biomineralization—unlike bone osteoblasts for example—yet it contains the machinery to undertake such a complex construction process. We might therefore expect to find some general principles in this example that help us lay out a conceptual framework to guide our thinking about other biomineralization systems.

The first stage of coccolith formation involves the assembly and organization of a series of membrane-bounded vesicles on the rim of an organic baseplate prior to mineralization. We can equate this with a general principle of *supramolecular preorganization*, which serves as a mechanism for establishing a 'building site' prior to the arrival of the bricks and mortar. This principle was implicit in much of our discussion in Chapter 5 on boundary-organized biomineralization, and as shown in Table 8.1, applies to various types of supramolecular assemblies, including, for example, the organized structure of collagen fibrils.

Once the vesicles are established around the rim of the organic baseplate, the second stage of coccolith construction begins. This involves the oriented nucleation of one calcite crystal per vesicle, and again we can identify this with a general principle—in this case *interfacial molecular recognition*—in which the preorganized supramolecular assemblies are encoded with sufficient information to direct the nuclei along certain crystallographic directions. Chapter 6, Section 6.7 described some of the possible types of interactions that could account for this remarkable degree of molecular specificity at the surface of an organic matrix. After nucleation, a third level of construction is brought into play, which in the coccoliths involves the growth of the immature crystals along predetermined directions to produce single-domain crystals with complex form. The overarching principle here is *vectorial regulation* and embraces the types of patterning mechanisms we discussed in Chapter 7. Finally, the coccolith plates are moved to the cell wall and used as prefabricated building blocks for the *higher-order assembly* of the coccosphere. From a wider perspective, this is analogous to the development of hierarchy in bone microstructures (Section 8.1), in which a series of related osteon-based building blocks become embedded as the length scale increases.

Table 8.1 Supramolecular assemblies in biomineralization

Assembly	Constituents	System
Nanoaggregates	Amelogenins	Enamel
	Ice-nucleating proteins	Bacterial cell walls
Nanocapsules	Ferritin proteins	Mammals, bacteria
Vesicles	Phospholipids	Magnetic bacteria
		Diatoms, etc.
2-D Templates	Surface layer proteins	Bacterial cell walls
	β-Sheet glycoproteins	Shells
3-D Crystalline nets	Collagen	Bone
3-D Macromolecular nets	α/β-Chitin	Limpet teeth
		Crab cuticle
	Cellulose	Plant walls
	Mucopolysaccharides	Bacterial cell walls

In summary, and as a general model for further development, we can consider biomineral tectonics to involve a multilevel process that contains at least four constructional stages:

Stage 1 Supramolecular preorganization
↓
Stage 2 Interfacial molecular recognition
↓
Stage 3 Vectorial regulation
↓
Stage 4 Higher-order assembly

It might help (it might not!) to think of this in terms of a commonplace analogy, such as building a house. First, the building plot or site is established. This means clearing a space and setting well-defined boundaries so that the size, shape and connectivity of the individual plot are agreed on before commencing the building programme (stage 1). Also, access to the site has to be secured so that materials can be supplied and removed at later stages. Next, the foundations have to be laid using a plan or blueprint that has sufficient information to dictate the basic pattern for construction of the dwelling (stage 2). Once this is achieved, the house can now be fabricated in line with predefined specifications such that bay windows, gable roofs, etc., are positioned in their correct locations (stage 3). If a series of building sites are established within a location, the result is not only the predetermined construction of individual houses but also the higher-order organization of houses into a street (stage 4). In terms of hierarchy, streets are embedded within housing estates, housing estates within towns, and towns within cities, etc.

There are four sequential constructional stages in biomineral tectonics: supramolecular preorganization, interfacial molecular recognition, vectorial regulation and higher-order assembly.

8.5 Summary

The formation of biominerals with structures that transcend a single level of scale is a remarkable example of how biological processes have assimilated the chemistry of inorganic precipitation for the construction of complex organized materials. As a start towards understanding how these structures are constructed, this chapter has introduced the concept of biomineral tectonics, which is underpinned by the principle of multilevel processing. Using calcified coccoliths as an example, we have identified at least four stages of construction—supramolecular preorganization, interfacial molecular recognition, vectorial regulation and higher-order assembly—which have generic significance for biomineralization in unicellular and multicellular organisms.

Extending biomineralization across many length scales requires a series of coordinated processes and building blocks of increasing size. We discussed how this is achieved in bone by embedding organized microstructures of collagen and bone cells into a hierarchical programme of construction. In unicellular organisms, by comparison, organic matrices are uncommon and the increased length scale is achieved by using prefabricated mineral building blocks trapped inside vesicles, which are often translocated to extracellular sites that are in close proximity to the cell wall. Although the building units might be simply 'dumped' into the mineralization front, they are often assembled into organized structures.

This chapter completes our overview of the principles and concepts of biomineralization. Together with the concepts described in Chapters 4 to 7, biomineral tectonics provides a framework for the development of a *biomineral-inspired* approach to materials chemistry. This is the subject of Chapter 9.

Further reading

Didymus, J. M., Young, J. R., and Mann, S. (1994). Construction and morphogenesis of the chiral ultrastructure of coccoliths from the marine alga *Emiliania huxleyi*. *Proc. R. Soc. London B*, **258**, 237–245.

Dopping-Hepenstal, P. J. C., Ali, S. J. and Stamp, T. C. B. (1981). Matrix vesicles in the osteoid of human bone. In *Matrix vesicles* (ed. Ascenzi, A., Bonucci, E. and de Barnard, B.), pp. 229–234. Wichtig Editoe, Milan.

Francillon-Viellot, H., de Buffrenil, V., Castanet, J., Gerandie, J. and Meunier, F. J. (1990). Microstructure and mineralization of vertebrate skeletal tissue. In *Skeletal biomineralization: patterns, processes and evolutionary trends* (ed. Carter, J. G.), pp. 471–530. Van Nostrand Reinhold, New York.

Hemleben, Ch., Anderson, O. R., Berthold, W. and Spindler, M. (1986). Calcification and chamber formation in foraminifera—a brief overview. In *Biomineralization in lower plants and animals* (ed. Leadbeater, B. S. C. and Riding, R.), pp. 237–249. Systematics Association Vol. 30. Oxford University Press, Oxford.

Leadbeater, B. S. C. (1979). Developmental studies on the loricate choanoflagellate *Stephanoeca diplocostata* Ellis. II. Cell division and lorica assembly. *Protoplasma*, **98**, 311–328.

Leadbeater, B. S. C. (1986). Silica deposition and lorica assembly in choanoflagellates. In *Biomineralization in lower plants and animals* (ed. Leadbeater, B. S. C. and Riding, R.), pp. 345–359. Systematics Association Vol. 30. Oxford University Press, Oxford.

Wainwright, S. A., Biggs, W. D., Currey, J. D. and Gosline, J. M. (1976). *Mechanical design in organisms*. Princeton University Press, Princeton, NJ.

Young, J. R., Davis, S. A., Bown, P. R. and Mann, S. (1999). Coccolith ultrastructure and biomineralisation. *J. Struct. Biol.*, **12**, 195–215.

Young, J. R., Didymus, J. M., Bown, P. R., Prins, B. and Mann, S. (1992). Crystal assembly and phylogenetic evolution in heterococcoliths. *Nature*, **356**, 516–518.

9 Biomineral-inspired materials chemistry

The study of biomineralization offers valuable insights into the scope of materials chemistry at the inorganic–organic interface. We have described how biominerals are highly controlled in structure, composition, shape and organization, and these properties are directly related to many areas of materials chemistry such as

- inorganic–organic composites
- nanomaterials
- functional materials and interfaces
- oriented crystals
- materials with complex morphologies
- organized assemblies
- hierarchical materials.

The route from biomineralization to materials chemistry involves *biomimetics*. There are several approaches that this can take, and indeed a whole book could be written on this subject alone (see further reading). Because this is a book about chemistry, our approach is focused on how an understanding of biomineralization has inspired new strategies in materials chemistry *synthesis*. Other aspects of biomimetics, such as the adaptation of biomechanical design in the development of novel materials, are not addressed.

There are three main areas of biomineralization that are currently being developed in biomimetic materials synthesis:

- biological concepts
- biological molecules and matrices
- biological systems.

This chapter is primarily concerned with how concepts in biomineralization can be translated into methods and strategies for use in the laboratory synthesis of materials. Sometimes this involves the use of proteins (Section 9.2.2) or polysaccharides (Section 9.3.1) extracted from biominerals, and occasionally intact living systems are employed (Section 9.2.3). But in the majority of cases, synthetic analogues are used to broaden the scope of the biomimetic approach.

Biomineralization processes are archetypes of synthetic strategies for the construction of organized materials across a range of length scales. The application of biomineralization concepts, molecules and systems is inspiring a biomimetic approach in inorganic materials chemistry.

9.1 Concepts and strategies

Underpinning the interrelationship between biomineralization and materials chemistry is the general notion that complex materials can be constructed from chemically based processes that extend across various length scales and which are integrated into higher levels of organization. In previous chapters we have discussed several principles that help to understand how this comes about in biomineralization, and these can be used as starting points for the development of analogous strategies in inorganic materials synthesis (Table 9.1). For example, the use of supramolecular assemblies prior to biomineralization (Chapter 5) is inspiring ideas based on the *spatial confinement* of chemical reactions and their materials products. Similarly, the concept of interfacial molecular recognition (Chapter 6) is motivating many scientists to search for analogous synthetic systems in which functionalized organic surfaces are used in the *template-directed* control of nucleation and architecture on various length scales. Likewise, the vectorial regulation of biomineral morphogenesis and pattern formation (Chapter 7) is stimulating approaches to the *morphosynthesis* of inorganic materials with complex form. And notions of multilevel processing in biomineral tectonics (Chapter 8) are related to the development of strategies for the long-range chemical construction of organized architectures from preformed crystalline building blocks such as inorganic nanoparticles. We shall refer to this approach as *crystal tectonics*.

Several examples of these biomineral-inspired approaches are discussed in the following sections of this chapter. References to the original papers, along with review articles, can be found at the end of the chapter.

Biomineralization concepts, such as supramolecular preorganization, interfacial molecular recognition, vectorial regulation and multilevel processing, are inspiring biomimetic approaches to inorganic materials synthesis based on spatial confinement, organic templates, complex form and multilevel processing, respectively.

9.2 Synthesis in confined reaction spaces

The ability of supramolecular compartments such as phospholipid vesicles to control the spatial dimensions of many biominerals suggests that analogous systems should be available and exploitable in synthetic materials chemistry across a range of length scales. For example, in Chapter 5 we discussed several boundary-organized systems—vesicles, ferritin, cellular assemblies

Table 9.1 Biomineralization concepts and related biomimetic approaches in inorganic materials synthesis

Process	Concept	Synthetic strategy
Boundary-organized biomineralization	Supramolecular preorganization	Confinement
Organic matrix-mediated biomineralization	Interfacial molecular recognition	Template-directed
Morphogenesis	Vectorial regulation	Morphosynthesis
Biomineral tectonics	Multilevel processing	Crystal tectonics

Table 9.2 Organic boundaries in biomineralization and their biomimetic counterparts in spatially confined materials synthesis

Biomineralization	Biomimetic synthesis
Phospholipid vesicles	Synthetic vesicles
Ferritin	Artificial ferritins
Cellular assemblies	Bacterial threads
Macromolecular frameworks	Polymer sponges

and macromolecular frameworks—and each of these has a direct synthetic counterpart, as shown in Table 9.2.

We begin with some examples that involve biomineral-inspired approaches to the synthesis of inorganic nanoparticles using boundary-organized nanoscale reaction droplets (Table 9.3). Nanoparticle synthesis is particularly important in materials chemistry because as inorganic particles become smaller, their electronic, optical and chemical properties begin to change. For example, in semiconductor materials such as CdS, quantum-size effects give rise to increases in the UV–visible absorption edge energy that can be used to tune the photoresponsive behaviour of these materials. Nanoparticles are also used extensively in heterogeneous catalysis because of their large surface areas.

Synthesis in confined spaces requires the supramolecular assembly of structures that contain chemical reactions or phase transformations. Although a range of length scales can be envisaged, the biomimetic materials synthesis of inorganic nanoparticles is particularly important.

Table 9.3 Biomineral-inspired approaches to the synthesis of inorganic nanoparticles and composites in confined spaces

Approach	Product	System	Materials
Boundary-organized reaction spaces	Surfactant-coated clusters	Reverse micelles	CdS, $BaSO_4$
		Microemulsions	Pt, Co, metal borides
			Fe_3O_4, $CaCO_3$
	Membrane-bounded nanoparticles	Vesicles	Pt, Ag, CdS, ZnS
			Ag_2O, FeOOH, Fe_3O_4, Al_2O_3
			Ca phosphates
	Artificial proteins	Ferritin	MnOOH, UO_3, FeS, Fe_3O_4
			CdS
		Viroid cages	Tungstates
	2-D nanoparticle superlattices	Porous S-layers	Ta/W, CdS, Au
Internally organized extended structures	Mineral–organic mesostructures	Lipid bilayer films	CdS, Fe_3O_4
		Multilamellar vesicles	SiO_2
	Bacteria–mineral fibres	Bacterial threads	SiO_2, zeolites
	Polymer–mineral composites	Copolymer sponges	Fe_3O_4, TiO_2
		Polyethylene oxide gels	CdS
		Collagen gels	Ca phosphates

9.2.1 Synthetic vesicles

Biological vesicles are central to many biomineralization processes—for example, see Chapter 3, Section 3.3.1, and Chapter 5, Section 5.1.1. In the laboratory, synthetic vesicles can be readily prepared by sonicating aqueous dispersions of phospholipid molecules, such as *phosphatidyl choline* (Fig. 9.1). Certain synthetic surfactants can also be used so the scope of this procedure is quite large. Before sonication, the amphiphilic molecules self-assemble into *multilamellar* vesicles that have an 'onion-like' mesostructure with possibly hundreds of concentrically arranged spherical bilayer membranes of increasing diameter separated by thin shells of water (Fig. 9.2). The high power of the ultrasound breaks up the bilayer shells, and the released fragments then curl up and reseal as *unilamellar* vesicles with a single aqueous-filled inner compartment, as shown in Fig. 9.2. In so doing, some of the ions and molecules present in the solution are encapsulated within the vesicles. By removing residual ions or molecules that remain outside the vesicle—column chromatography is often used—the vesicles can be exploited as confined reaction spaces provided that the trapped constituents don't leak out and a membrane-permeable coreactant is available. For example, a range of metal oxide and sulfide nanoparticles (see Table 9.3) can be prepared by reacting the encapsulated metal ions with OH^- and H_2S,

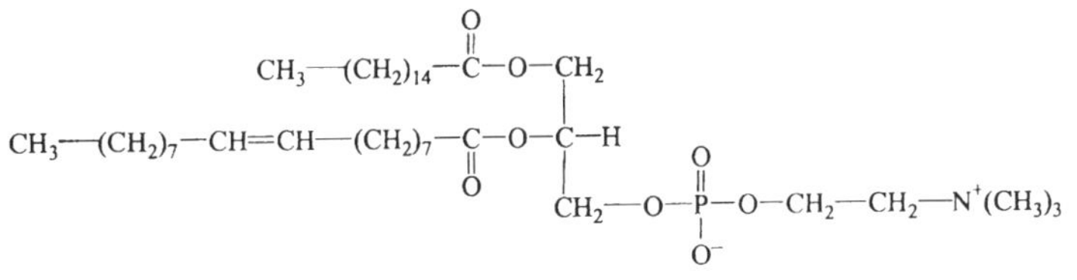

$CH_3-(CH_2)_{14}-C(=O)-O-CH_2$

$CH_3-(CH_2)_7-CH=CH-(CH_2)_7-C(=O)-O-C-H$

$CH_2-O-P(=O)(O^-)-O-CH_2-CH_2-N^+(CH_3)_3$

Fig. 9.1 Phosphatidyl choline.

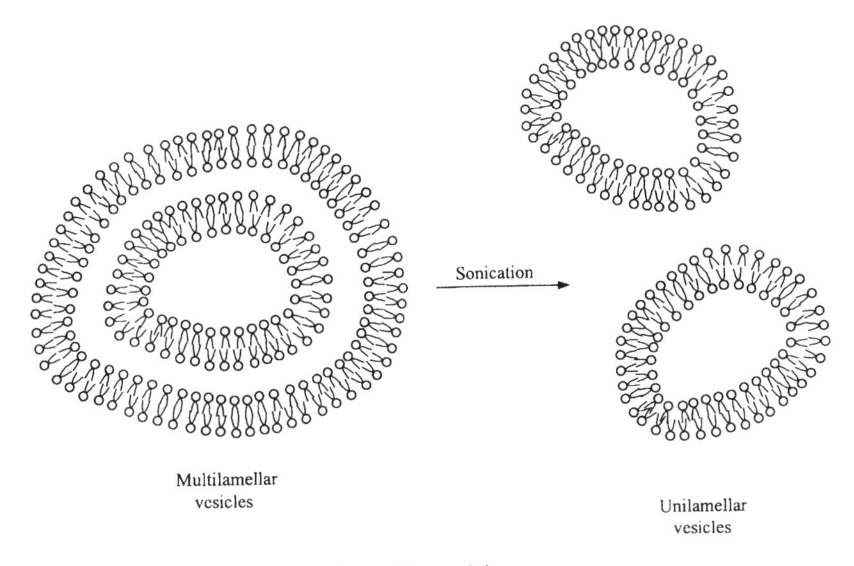

Fig. 9.2 Multilamellar and unilamellar vesicles.

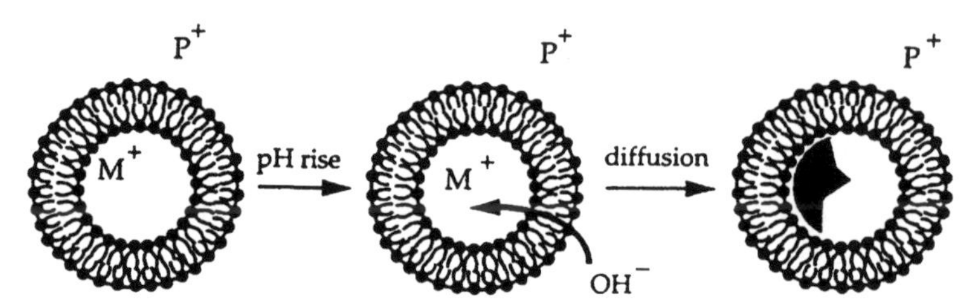

Fig. 9.3 Formation of metal oxide nanoparticles within unilamellar vesicles. Metal cations (M^+) are encapsulated and reacted with OH^- ions that diffuse through the bilayer membrane. Inert cations (P^+) are required to maintain electroneutrality.

respectively, which readily diffuse through the bilayer membrane, as shown in Fig. 9.3. The size of the nanoparticles is determined by the number of metal ions encapsulated within the interior of the unilamellar vesicles. In most cases, this gives rise to spherical particles usually less than 20 nm in diameter. And because a lipid bilayer membrane surrounds each inorganic particle, they do not coalesce or fuse into larger aggregates (Fig. 9.4).

Precipitation within the vesicles can be monitored by the technique of light scattering. For Ag_2O, there is a linear (first-order) relationship between the initial rate of precipitation and the concentration of Ag^+ inside the vesicle ($[Ag^+]_{in}$) at constant pH. A two-step reaction mechanism has been proposed:

(1) $$[OH^-]_{out} \rightarrow [OH^-]_{in}$$

(2) $$2[OH^-]_{in} + 2[Ag^+]_{in} \rightarrow [Ag_2O]_{in} + H_2O$$

The first step is diffusion-controlled and depends on the rate of passage of OH^- ions through the lipid membrane. At pH_{out} values below 10 the rate of OH^- influx is very small due to the small concentration gradient, and supersaturation is never attained within the vesicles. Between pH values of 11 to 12, however, the reaction is strongly dependent on $[OH^-]_{out}$ but above 12, the limiting rate of membrane diffusion is attained and intravesicular precipitation becomes independent of pH and is controlled by step 2.

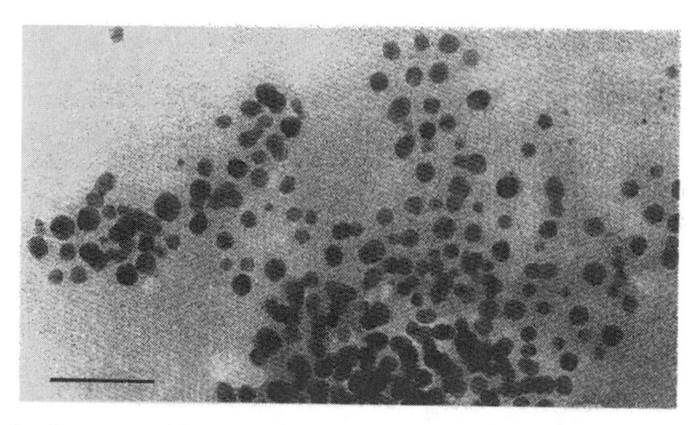

Fig. 9.4 Ag_2O nanoparticles prepared inside phosphatidyl choline unilamellar vesicles. Scale bar, 75 nm.

The relationship between the intra- and extravesicular pH is dependent on the presence of other anions inside the vesicles. In the case of $AgNO_3$ solutions, the intravesicular NO_3^- ions diffuse out of the vesicles as OH^- ions enter. This preserves the electroneutrality of the system and allows further influx of the hydroxide coreactant. Even so, a pH gradient of approximately 3 units across the vesicle membrane has to be established before this exchange process takes place. In contrast, if NO_3^- ions inside the vesicles are replaced with an impermeable doubly charged anion such as SO_4^{2-}, then no OH^- diffusion occurs and pH gradients of up to 6 units can be achieved across the lipid membrane without precipitation of Ag_2O. Similarly, if Cl^- is substituted for OH^- then the rate of intravesicular AgCl precipitation from aqueous $AgNO_3$ is significantly reduced because of the lower membrane permeability of the coreactant. The formation of intravesicular Ag_2S, on the other hand, occurs instantaneously because H_2S molecules rapidly diffuse across the phospholipid membrane.

Multilamellar vesicles can be used to prepare organized *nanostructures* by mineralization of the regularly spaced aqueous-filled interlayer regions. For example, the preparation of multilamellar vesicles with the protonated surfactant 1,12-diaminododecane ($^+H_3N(CH_2)_{12}NH_3^+$) dispersed in water containing tetraethoxysilane ($Si(OCH_2CH_3)_4$, TEOS) results in the formation of thin concentric layers of silica by hydrolysis and condensation reactions located specifically within the aqueous interlayer spaces (Fig. 9.5). Each vesicle gives rise to a spherical micrometre-sized inorganic particle with an internal mesostructure consisting of multiple shells of a laminated silica–surfactant composite. Surprisingly, the surfactant template can be removed to give a porous silica replica without loss of structural integrity. A lamellar structure should collapse after removal of the surfactant bilayers so tiny pillars of silica must traverse the layers to maintain the porosity. One possibility is that the silica nanocolumns are formed from hydrolysed TEOS molecules that infiltrate membrane defects in the multilamellar structure.

Nanoparticle synthesis inside unilamellar phospholipid vesicles is diffusion-controlled and dependent on electrochemical gradients across the membrane. Infiltration and mineralization between the concentrically

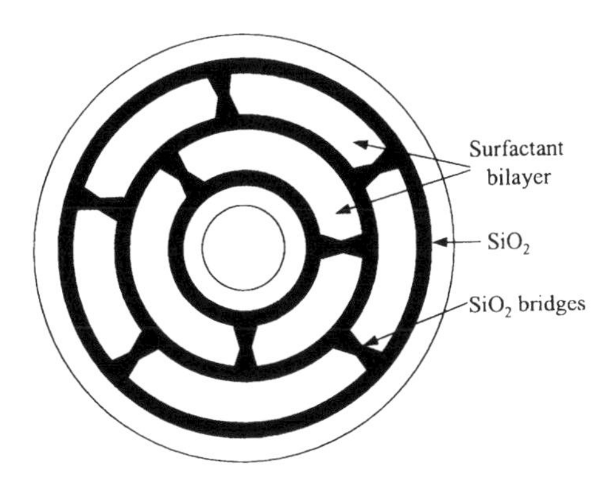

Fig. 9.5 Silica mineralization in multilamellar vesicles and formation of an onion-like inorganic–organic nanocomposite.

arranged bilayers of multilamellar vesicles produce inorganic particles with onion-like internal structure.

9.2.2 Artificial ferritins

The supramolecular structure of ferritin (see Chapter 5, Section 5.1.2) is remarkably stable for a biological molecule. Indeed, as we described in Chapter 6, Section 6.7.3, the iron oxide (ferrihydrite) cores can be chemically removed in the laboratory by reductive dissolution without affecting the structure of the protein shell. Moreover, the resulting empty cages of the demineralized protein (*apoferritin*) can be subsequently *reconstituted* under controlled experimental conditions involving incubation of the protein with aqueous Fe^{II} solutions in the presence of an oxidizing agent such as oxygen in the air. These unique features, along with the fact that ferritin is easily isolated from biological tissues and commercially available, provide opportunities to use this biomolecule in the spatially confined synthesis of protein-encapsulated nanoparticles with potential biocompatible properties (Table 9.3). *Artificial ferritins* are synthesized by two chemical methods (Fig. 9.6):

- *in situ* transformation of the native iron oxide cores
- *in situ* synthesis of non-natural materials.

The first approach has been used to prepare amorphous iron sulfide nanoparticles within the protein cage by chemical transformation of the iron oxide cores under an argon atmosphere in the presence of aqueous Na_2S. The reaction solution quickly turns from brown to dark green with new absorption maxima at 426 nm and 338 nm that are characteristic of iron–sulfur clusters. Because the corresponding nanoparticles are similar in dimension to the original iron oxide cores, the iron sulfide phase can be controlled in size from 2 to 7 nm by using ferritins reconstituted with predetermined loadings of iron oxide.

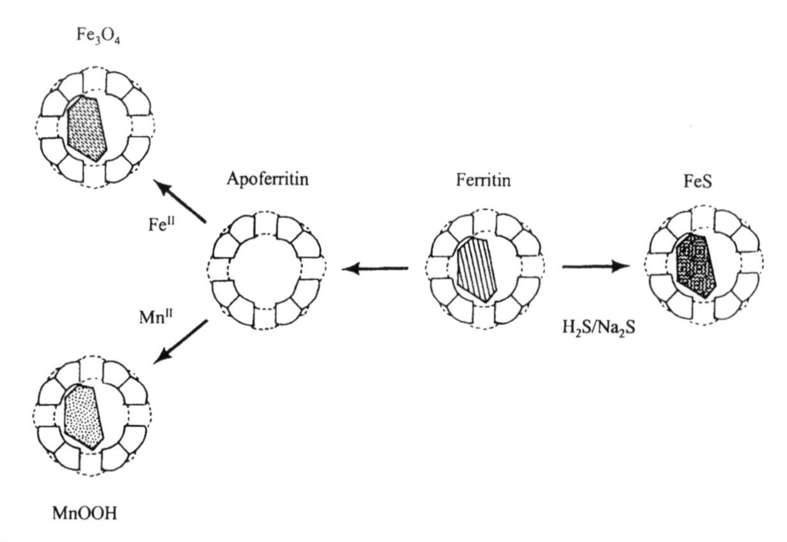

Fig. 9.6 Biomimetic synthesis of artificial ferritins.

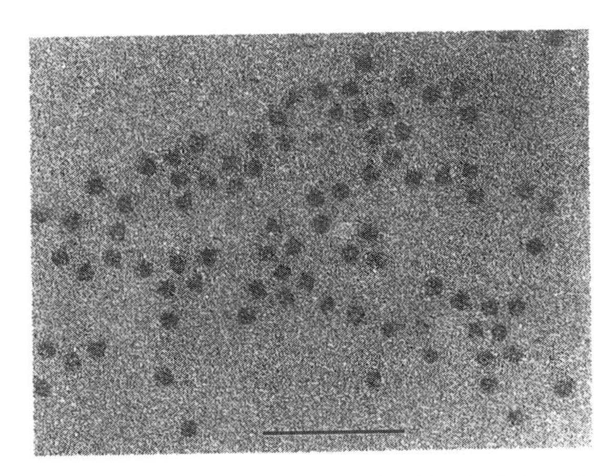

Fig. 9.7 MnOOH nanoparticles prepared in ferritin. Scale bar, 50 nm.

Demineralized ferritin has been used in the synthesis of a range of non-iron-containing nanoparticles, such as MnOOH (see Fig. 9.6), as well as CdS and hydrated uranium oxides (Table 9.3). Incubation of apoferritin with Mn^{II} solutions results in the uptake of Mn^{II} ions and subsequent aerial oxidation and deposition of discrete amorphous MnOOH cores in the polypeptide cage (Fig. 9.7). The size of the nanoparticles is controlled to some extent by changing the stoichiometry of the reaction mixtures. Mixed metal oxide cores, such as FeOOH/MnOOH nanoparticles, can also be prepared by alternating the addition of aqueous Fe^{II} or Mn^{II} ions to the protein solution.

The reconstitution approach is also a viable method to prepare a magnetic ferritin, called *magnetoferritin* (Fig. 9.6). Whereas natural ferritin is superparamagnetic and becomes weakly magnetic only at low temperatures (see Chapter 2, Section 2.6.2), the corresponding artificial protein prepared by anaerobic incubation of apoferritin with Fe^{II} solutions containing sub-stoichiometric amounts of an oxidant responds to a magnetic field even at room temperature. This is because the chemical conditions—elevated temperature (85°C), pH (8.5) and incomplete oxidation—give rise to the formation of the mixed valance phase, magnetite (Fe_3O_4), rather than the fully oxidized mineral, ferrihydrite, inside the polypeptide cavity. As shown in Fig. 9.8, the magnetization curves for magnetoferritin measured at 300 K exhibit saturation behaviour. This means that as the applied magnetic field is increased across a solution of magnetoferritin, the magnetite cores become progressively aligned and the induced magnet moment increases until a saturation value is attained where all the tiny compass needles are oriented along the direction of the external field. On reversing the direction of the field the cores realign without any hysteresis because there is only a very small activation energy barrier to be overcome for particles of this size. The situation is different at low temperature (20 K) where the magnetic dipole moment is not scrambled so readily by the thermal energy and a barrier therefore exists.

Artificial ferritins with a range of nanoparticle cores are produced by in situ chemical synthesis or phase transformation.

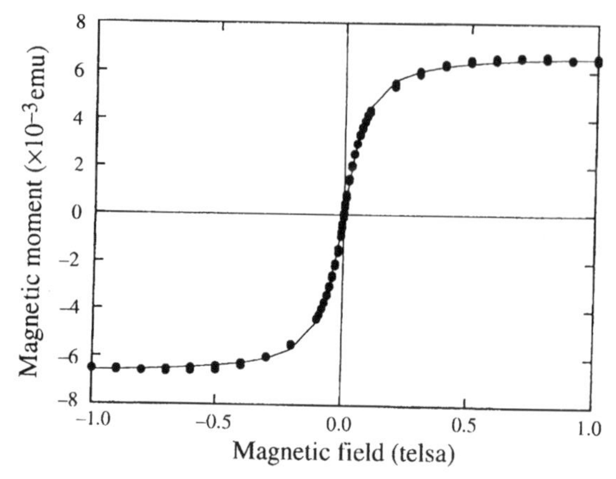

Fig. 9.8 Magnetization curve for magnetoferritin at 300 K.

9.2.3 Bacterial threads

In Section 5.1.3 of Chapter 5 we described how groups of cells are organized to seal off spaces for biomineralization. Could analogous systems be exploited in biomimetic materials synthesis?

Bacillus subtilis is a friendly bacterium that helps to make unfermented soybean curd, generally known as 'tofu'. It can be grown in culture in the form of long multicellular strings, one cell across and thousands of cells long. These 0.5-μm-wide filaments can be organized into a close-packed superstructure by dipping a small hook into the culture and slowly withdrawing it back through the air–water interface and beyond. With practice, a macroscopic fibre a few millimetres across and tens of centimetres in length can be produced. Held in the hand, the thread looks like a length of nylon fishing wire but in the scanning electron microscope a remarkable internal superstructure is revealed. This consists of a well-ordered hexagonal array of multicellular filaments that run parallel to the long axis of the macroscopic fibre (Fig. 9.9).

Fig. 9.9 Sectioned bacterial thread showing internal hexagonal superstructure of coaligned multicellular filaments. Scale bar, 10 μm.

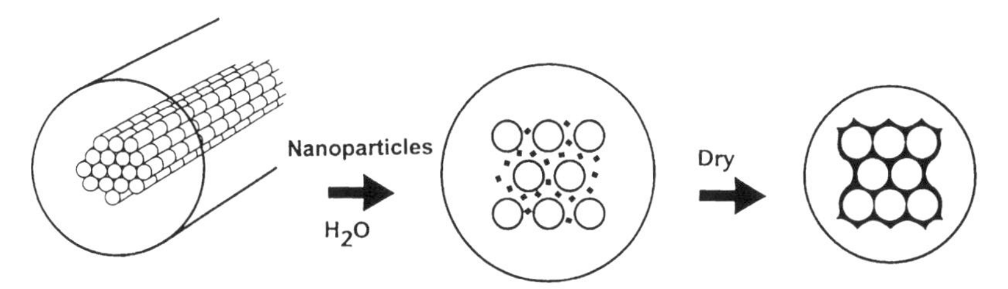

Fig. 9.10 Mineralization of bacterial superstructure using inorganic nanoparticles and reversible swelling.

The bacterial thread swells without major damage to the underlying superstructure when placed in water due to partial separation of the multicellular filaments as they become rehydrated. As a result, the hydrated structure consists of an organized array of aqueous channels that are delineated by the multicellular filaments. If the swelling procedure takes place in a sol of silica or zeolite nanoparticles, then the spaces between the filaments are filled with mineral to produce a 'bacterial skeleton', referred to as a *bionite* (Fig. 9.10). In the case of silica colloids, the negatively charged nanoparticles penetrate deep within the threads because they are repelled from the anionic cell walls of the bacterial filaments. Positively charged colloids such as TiO_2, in contrast, are rapidly aggregated on the surface of the thread so that infiltration of the superstructure is negligible.

As illustrated in Fig. 9.10, the mineral-containing thread contracts on drying and the inorganic nanoparticles are trapped between the multicellular filaments. Large increases in ionic strength occur within the interfilament regions during drying, and this shields the surface charge on the silica nanoparticles such that a continuous mineral framework with 200-nm-thick walls is formed by aggregation and compaction of the particles around the cellular filaments (Fig. 9.11).

Bacterial threads with an organized internal superstructure are used to spatially control the deposition of silica nanoparticles.

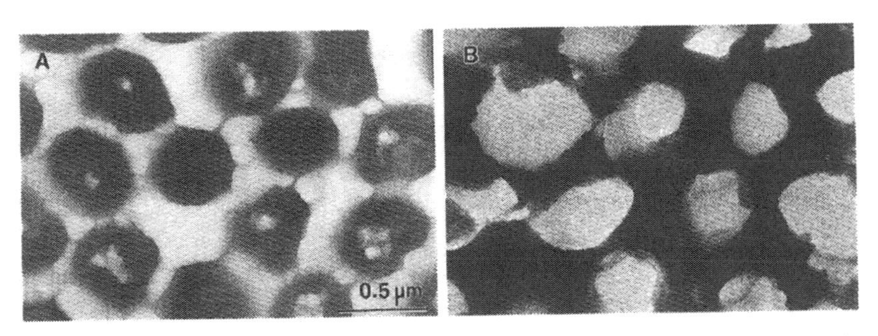

Fig. 9.11 Cross-section of bacterial thread: (A) before mineralization showing end-on view of the multicellular filaments and interfilament spaces; (B) after mineralization showing continuous silica walls between the entrapped filaments. Scale bar, 0.5 μm in both micrographs.

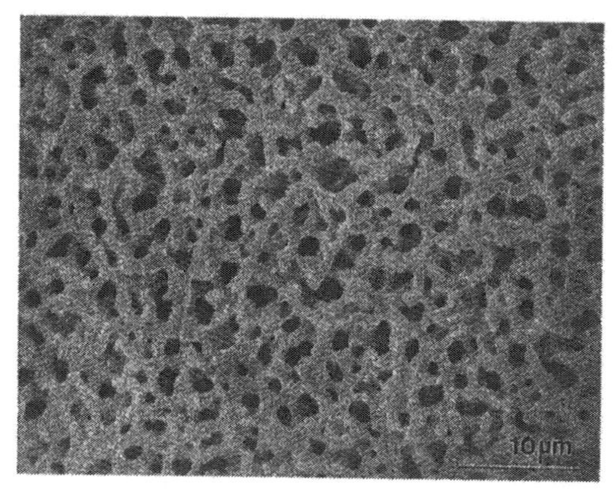

Fig. 9.12 Polymer sponge. Scale bar, 10 μm.

9.2.4 Polymer sponges

One of the key roles for organic macromolecular frameworks in biomineralization is to partition space for the construction of large structures such as eggshells, limpet teeth and mollusc shells (see Chapter 5, Section 5.1.4). A corresponding biomimetic counterpart of this process involves the use of synthetic organic polymer gels with sponge-like internal structure (see Fig. 9.12) that are infiltrated with reaction solutions or colloidal suspensions.

Polymer sponges can be prepared from mixtures of monomers such as acrylic acid with either styrene or 2-hydroxyethyl methacrylate (Fig. 9.13). When dry, the gel monoliths contract in size, but on contact with water they undergo reversible swelling to give an open-cell structure with pores several nanometres across that can be infiltrated and mineralized with inorganic components. The degree of swelling depends on the electrostatic repulsion between ionizable groups and therefore on the carboxylic acid content of the copolymers. Thus, by fine-tuning the polymer chemistry, different sponge-like architectures can be synthesized. For example, swelling a polystyrene–polyacrylate copolymer gel in an aqueous solution of Fe^{II} and Fe^{III} ions, followed by addition of NaOH, results in the *in situ* deposition of superparamagnetic magnetite (Fe_3O_4) nanoparticles throughout the open-cell framework. The mineralized composite has a structure that loosely resembles the arrangement of magnetite crystals within the polysaccharide matrix of chiton teeth (see Chapter 2, Section 2.6.3). Unlike the mollusc teeth, however, the synthetic material has magnetite loadings of only 8 per cent by weight because the supersaturation levels attainable within the template are low and cannot be sustained for long periods. One possible way round the problem is to use larger building blocks, such as preformed magnetite nanoparticles, to infiltrate the reversibly swollen polymer sponge. High levels of swelling are required for this, so spongy monoliths formed from copolymers containing 20 per cent polyacrylate and 80 per cent poly(2-hydroxyethyl methacrylate) are used. The resulting highly swollen, black polymer–inorganic composite is then dried in air to give a solid, hard and compact material that is readily attracted to the poles of a permanent magnet.

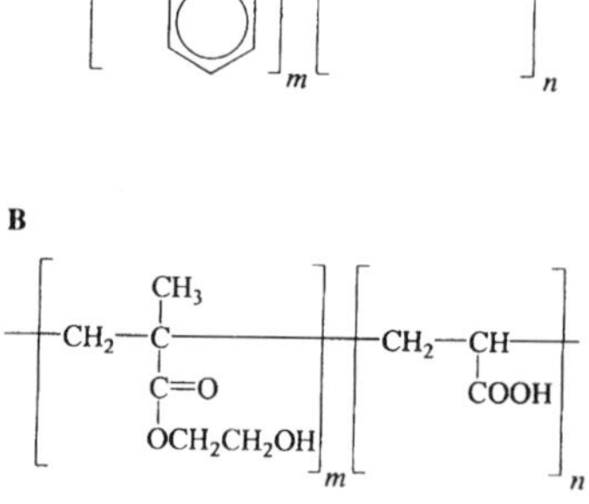

A

$$\left[\left[-CH_2-CH(C_6H_5)- \right]_m \left[-CH_2-CH(COOH)- \right] \right]_n$$

B

$$\left[\left[-CH_2-C(CH_3)(C{=}O\,OCH_2CH_2OH)- \right]_m \left[-CH_2-CH(COOH)- \right] \right]_n$$

Fig. 9.13 Copolymers: (A) styrene and acrylic acid; (B) 2-hydroxyethyl methacrylate and acrylic acid.

The internal spaces of polymer sponges can be infiltrated with minerals such as magnetite to produce organized inorganic–organic composites.

9.3 Template-directed materials synthesis

In Section 9.2 we described how preformed organic structures can be used to spatially confine inorganic deposition. In this section, we discuss several examples in which the interfacial properties of organic architectures are used to template the nucleation and growth of inorganic minerals.

Controlling the nucleation of inorganic solids is an important consideration in materials and colloid chemistry. Oriented inorganic substrates such as Au and Si are often used as epitaxial surfaces (see Chapter 4, Section 4.5) in methods involving chemical vapour deposition, and it would be significant if similar nucleation processes could be routinely undertaken in aqueous media. In Chapter 6 (Section 6.7) we described how molecular recognition at the surface of an organic matrix was of central importance in the controlled nucleation of biominerals. For this, structural and functional properties are combined so that both the spatial and structural aspects of mineralization are regulated. Synthetic analogues of organic matrix-mediated biomineralization are therefore based on self-assembled surfactant or polymeric templates that contain appropriate arrangements of surface functional groups. These interact with ions in the surrounding supersaturated solution to produce interfaces that resemble the first stages of nucleation so that deposition occurs specifically along the organic surface.

Some examples of this 'soft templates for hard materials' approach that have been inspired by biomineralization are shown in Table 9.4. Many of the systems involve the assembly of discrete supramolecular objects, such as organic tubes and filaments, and their subsequent surface mineralization. If the surface functional groups are arranged with some degree of geometric

Table 9.4 Biomineral-inspired approaches to the organic template-directed synthesis of inorganic materials

Approach	Product	System	Materials
Nucleation on biomineral matrices	Organized composites	β-Chitin/acidic macromolecules	$CaCO_3$, Ca phosphates
		Cuttlebone β-chitin	SiO_2
Nucleation on 3-D structures	Mineral–organic cylinders/tubes	Lipid tubules	Cu, Ni, Al_2O_3, Fe oxides, Au
		Viroid tubules	CdS, PbS, SiO_2, Fe oxides
		Bacterial rhapidosomes	Pd
		Bacterial fibres	$CaCO_3$, CuCl, Fe oxides
Nucleation on thin films	Oriented crystals	Langmuir monolayers	NaCl, $CaCO_3$, $BaSO_4$, PbS $CaSO_4$, $CuSO_4$
		Polyaspartate/polystyrene	$CaCO_3$
	Surface coatings	Self-assembled monolayers	TiO_2, zeolites
		Polyacrylate films	Fe oxides, $BaTiO_3$

periodicity or crystal face-specific complementarity, for example in a compressed monolayer of surfactant molecules, then as in biomineralization oriented nucleation is sometimes observed (see Section 9.3.3). In cases where the template is packed together into an extended structure, such as an array of filaments or stack of membranes, then the organic structure not only is surface-active but also plays an important role in the spatial confinement of the mineralization process, as described in Section 9.2.

Some examples of template-directed materials synthesis are discussed below.

Template-directed materials synthesis involves the specific nucleation and growth of inorganic phases on the surface of functionalized organic structures. Oriented nucleation can occur if there is a high level of molecular recognition at the template–mineral interface.

9.3.1 Biomineral matrices

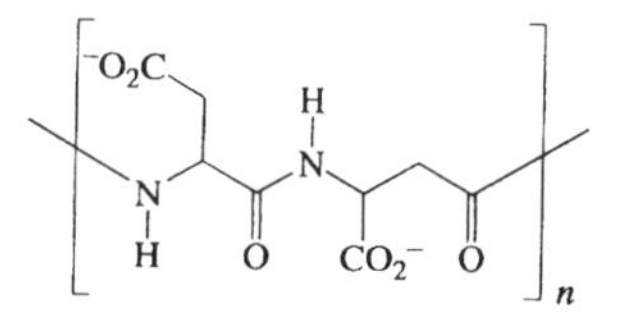

Fig. 9.14 Polyaspartate. Note the presence of two types of linkages, referred to from left to right as α and β, which depend on the position of the carboxylate side chain.

One possible approach to template-directed materials synthesis involves the adaptation of organic matrices obtained from the demineralization of biological tissues such as shells, bone and teeth. Because water-soluble macromolecules are often lost during extraction of the matrix, the hydrophobic components have to be 'reactivated'. For example, sheets of β-chitin (see Chapter 6, Section 6.5 and Fig. 6.19) have no significant influence on calcium carbonate crystallization in the laboratory unless acidic macromolecules, such as soluble shell nacre proteins or synthetic analogues like polyaspartate (Fig. 9.14), are first bound to the hydrophobic sheets. Another strategy is to use a synthetic substrate such as polystyrene along with surface-adsorbed polyaspartate macromolecules (see further reading). Under certain circumstances, the above combinations give rise to the oriented nucleation of calcite and therefore can be used as experimental models for understanding the process of organic matrix-mediated nucleation in biomineralization, as described in Chapter 6, Section 6.7.

In a few cases, the hydrophobic matrix associated with a biomineralized structure can be extracted without significant disruption to the 3-D architecture, which can then be exploited as a framework for inorganic deposition. For example, demineralization of the aragonitic internal shell (*cuttlebone*) of the cuttlefish (see Chapter 7, Section 7.33 and Fig. 7.21) produces an intact sponge-like organic monolith that consists of pure β-chitin (Fig. 9.15). The matrix can be remineralized directly without adding more macromolecules if supersaturated solutions of silica rather than calcium carbonate are used. This is because the surface of the chitin matrix contains sufficient numbers of hydroxyl groups to promote the interfacial nucleation of silica from alkaline aqueous silicate solutions. Provided that the supersaturation levels are relatively low, high-fidelity silica replicas with chamber-like architecture are produced (Fig. 9.16). These artificial cuttlebones might be useful as high surface area membranes in catalysis and chromatography.

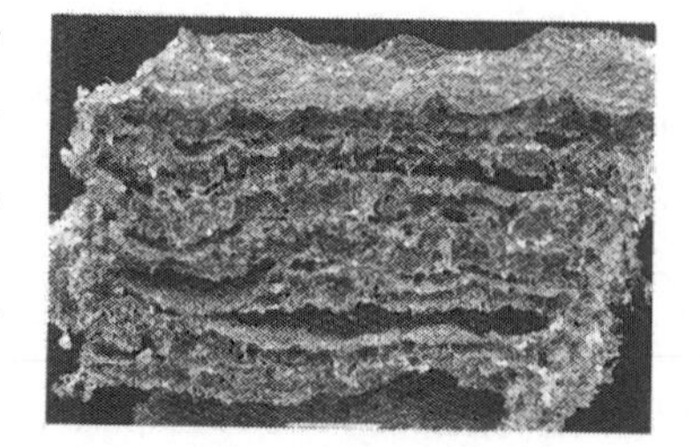

Fig. 9.15 Intact sponge-like β-chitin matrix from demineralized cuttlebone. Scale bar, 500 μm.

Hydrophobic organic matrices can be extracted from biomineralized structures and used as templates for the controlled synthesis of inorganic materials.

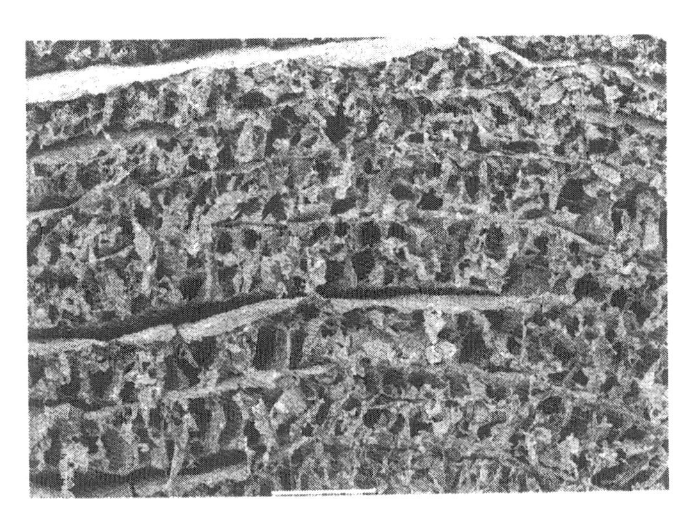

Fig. 9.16 Silica replica of the β-chitin matrix of cuttlebone. Scale bar, 500 μm.

9.3.2 Lipid tubules

Biomineral organic matrices are not easy to obtain, particularly in large amounts, so it makes sense to search for synthetic analogues with functionalized surfaces for inorganic nucleation. For example, several different types of tubular organic structures have been used as templates for the formation of filamentous inorganic–organic composites (Table 9.4). *Lipid tubules* are multilamellar structures formed from the supramolecular assembly of amphiphilic molecules that are chiral. In the initial stages of self-assembly, the molecules pack together in bilayer sheets separated by the solvent. But the molecular chirality gives rise to long strings of strongly-interacting chiral amphiphiles within each layer of the sheet structure and this produces stresses within the plane of the bilayer that cause the sheets to bend and roll up like a carpet.

Lipid tubules can be prepared from the biological lipid *galactocerebroside*, which consists of a sugar headgroup attached to a ceramide linker with various headgroup modifications including hydroxy (OH-Cer), non-hydroxy (H-Cer) and sulfated (S-Cer) derivatives (Fig. 9.17). The use of these tubules as templates for the synthesis of rod-like iron oxide composites is illustrated in Fig. 9.18. Whereas the sulfated and non-hydroxy derivatives do not spontaneously self-assemble into discrete tubules, the hydroxylated compound forms stable rolled-up multilamellar cylinders. But because the tubules are uncharged they are ineffective as templates for iron oxide precipitation unless a small amount (10 mol%) of the sulfated derivative is included in the lipid tubules. This has a marked effect on the nucleation specificity, and mineralized tubes several micrometres in length and 30 to 50 nm in width are produced by incubating suspensions of the tubules in Fe^{III}-containing solutions (route A in Fig. 9.18). The sulfated sugar headgroups act as metal-ion binding sites for Fe^{III}, which under appropriate solution conditions facilitate the specific nucleation and growth of a thin coating of lepidocrocite (γ-FeOOH) on the external surface of the lipid tubule (Fig. 9.19). As shown in route B of Fig. 9.18, *in situ* conversion of the lepidocrocite-coated tubules to a magnetic composite is achieved

OH
$(CH_2)_{12}CH_3$
HO
O
N
OH
Y
X
$(CH_2)_{20}CH_3$
O
H

X = OH	Y = OH	HO - Cer
X = H	Y = OH	H - Cer
X = OH, H	Y = OSO_3^-	S - Cer

Fig. 9.17 Molecular structure of galactocerebroside lipid and three derivatives.

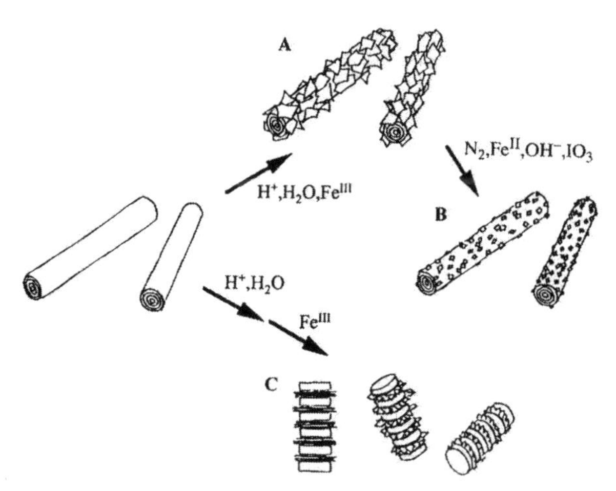

Fig. 9.18 Use of galactocerebroside lipid tubules in the template-directed synthesis of iron oxides. See text for details.

Fig. 9.19 Galactocerebroside lipid tubule coated with lepidocrocite (γ-FeOOH). Scale bar, 500 nm.

under N_2 at room temperature by adding aqueous Fe^{II} followed by an increase in the pH to 8.5 or an oxidant such as IO_3^-. Black magnetic tubes encrusted with clusters of fine-grained magnetite are formed after 4 days.

Very different results are obtained if the lipid tubule suspension is incubated for 10 minutes in acid solution *prior* to addition of Fe^{III} (route C in Fig. 9.18). The acid destabilizes the cylindrical template and 6.5-nm-thick discs of the sulfated S-Cer lipid are released into the solution by phase separation. Subsequent addition of aqueous Fe^{III} causes aggregation of the discs into stacks due to charge neutralization of the sulfated headgroups. With time, hydrolysis and condensation of the Fe^{III} ions to FeOOH occurs with the consequence that layers of lepidocrocite only a few nanometres wide are interspaced at regular distances along the stacked assembly of discs.

The external surfaces of supramolecular lipid tubules can be used to direct the nucleation and growth of inorganic minerals. Galactocerebroside tubules and discs containing sulfated headgroups are potent nucleators of iron oxide minerals.

9.3.3 Oriented nucleation on soap films

As described in Chapter 6 (Section 6.7.5), one of the most intriguing aspects of organic matrix-mediated biomineralization is oriented nucleation. This is thought to arise in part from structural matching between lattice spacings in certain crystal faces and distances that separate functional groups periodically arranged across the organic surface. Although there is only indirect evidence to support this hypothesis, there are some reasonable laboratory-based models to support the notion of structural recognition. Perhaps the best example of this is based on the controlled crystallization of inorganic solids on compressed monomolecular films of insoluble surfactants spread at the air–water interface. These thin soap films are referred to as *Langmuir monolayers*, and are easily produced by carefully placing a drop of a surfactant solution in chloroform onto the surface of water placed in an enclosed trough or tray. The drop rapidly spreads over the water surface and the chloroform evaporates to leave a monolayer of surfactant molecules that are oriented with their hydrocarbon chains approximately perpendicular to the liquid surface so that only the polar headgroups are hydrated. A physical barrier is moved across the surface of the water so that the surface area of the monolayer is gradually reduced and the molecules are squeezed together, which increases the surface pressure in the film. As the pressure rises, the molecules are marshalled from an expanded 'gaseous' state to a semi-ordered 'liquid' state and finally into an ordered 2-D crystalline 'solid' state in which a regular array of surfactant headgroups is exposed to the water (Fig. 9.20). Knowing how many surfactant molecules are added to the air–water interface and the area confined by the barrier allows the area per molecule to be calculated for any surface pressure. For surfactants with single alkyl chains and small headgroups the limiting area per molecule obtained for the compressed solid phase on pure water is in the range of 18 to 24 Å^2.

By replacing the water in the Langmuir trough with a supersaturated solution, and varying the degree of film compression and type of surfactant headgroups used, a wide range of experiments can be undertaken to test out ideas about oriented nucleation in biomineralization and how these can be adapted for template-directed materials synthesis. Table 9.4 gives some examples of

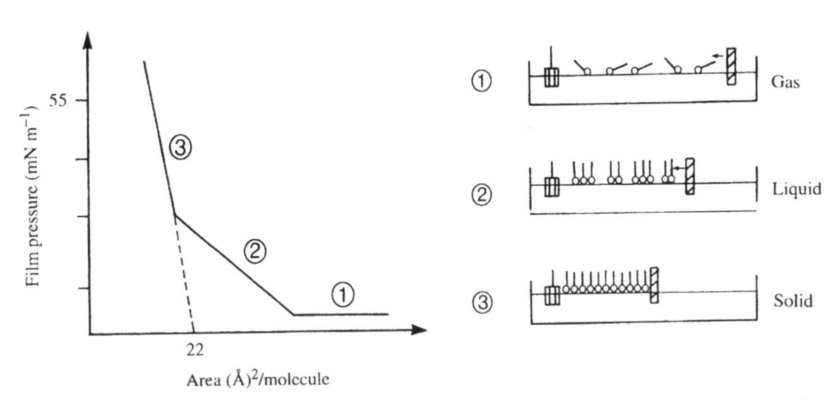

Fig. 9.20 Idealized plot of film pressure against area per molecule for a surfactant undergoing compression at the air–water interface. The corresponding gas, liquid and solid states of the monolayer are also shown.

the mineral phases studied. For $CaCO_3$, supersaturated calcium bicarbonate solutions are prepared at pH 6.0 by purging suspensions of calcite with CO_2 gas for 1 hour. This dissolves the calcite due to the increasing partial pressure of CO_2 in the system. When the gas flow is stopped and the solution poured into the Langmuir trough, slow outgassing of the CO_2 occurs and calcium carbonate crystallization is induced over several hours, according to the equilibrium,

$$Ca^{2+} + 2HCO_3^- \rightleftharpoons CaCO_3 + CO_2\uparrow + H_2O$$

This process is similar to that discussed for biological calcification in green algae that we described in Chapter 5, Section 5.4.1, and in the laboratory gives rise to aggregates of rhombohedral calcite crystals in the solution and at the air–water interface. But when a monolayer of *stearic acid* ($CH_3(CH_2)_{16}COOH$) is compressed to the solid state at the air–water surface, the mineral nucleates predominately at the air–water interface as discrete rhombohedral calcite crystals that are oriented with the $\{1\bar{1}0\}$ face parallel to the water surface (Fig. 9.21A). Amazingly, if a long alkyl chain sulfate—*n-eicosyl sulfate* ($CH_3(CH_2)_{19}OSO_3H$)—is used instead of stearic acid then triangular calcite crystals oriented with the $\{001\}$ face parallel to the monolayer–water interface are nucleated (Fig. 9.21B).

A structural model has been proposed to account for these observations. X-ray diffraction indicates that the negatively charged surfactant molecules are closely packed at the air–water interface in a distorted hexagonal unit cell with inter-headgroup distances of about 0.5 nm. A diffuse layer of Ca^{2+} ions forms under the monolayer by electrostatic attraction with the anionic headgroups but this is only structurally ordered when the metal-ion concentration in solution is relatively high, at about 10 mM. Under these conditions, oriented nucleation of either the $\{1\bar{1}0\}$ or the $\{001\}$ face can be rationalized because in both cases there is a good lattice match in two directions between the Ca^{2+} ions in the faces and Ca^{2+} ions bound to the anionic headgroups (Fig. 9.22). So what mechanism explains the 90° change in orientation when the

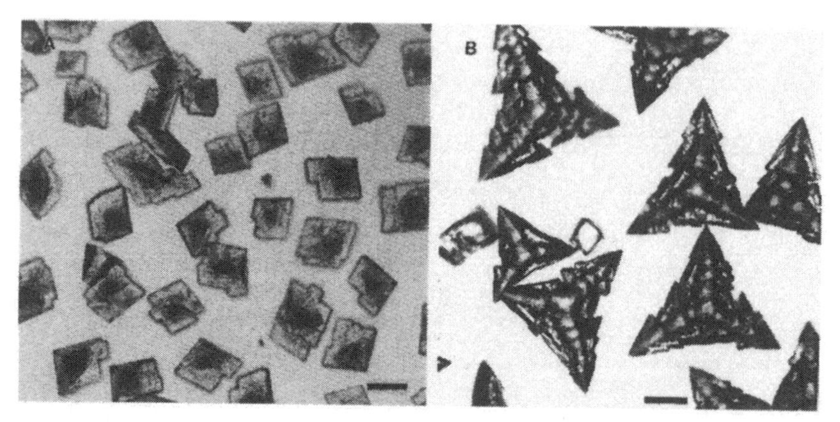

Fig. 9.21 Calcite crystallization under compressed Langmuir monolayers: (A) stearic acid film with $\{1\bar{1}0\}$ oriented crystals, scale bar, 50 μm; (B) *n*-eicosyl sulfate film with $\{001\}$ oriented crystals, scale bar, 20 μm.

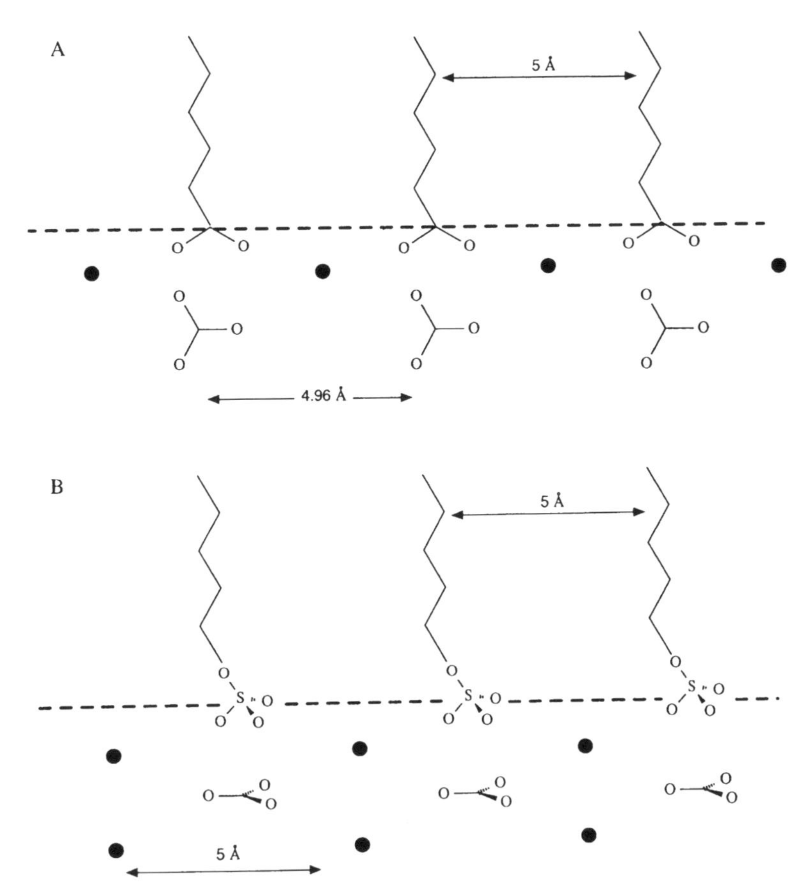

Fig. 9.22 Interfacial recognition under Langmuir monolayers and oriented calcite nucleation for: (A) carboxylate monolayers and the $\{1\bar{1}0\}$ face; (B) sulfate monolayers and the {001} face.

carboxylates are replaced with sulfate headgroups? The model proposes that along with the matching of lattice distances there is stereochemical recognition between the oxygen atoms of the surfactant headgroups and those in the carbonate anions in the crystal faces. For the $\{1\bar{1}0\}$ face, the carbonate groups are perpendicular to the crystal surface such that only two of the three oxygens are exposed (Fig. 9.22A), whereas carbonate anions lie parallel to the {001} face with three oxygens in the plane (Fig. 9.22B). These arrangements are mimicked, respectively, by the carboxylate and sulfate headgroups in the monolayer so that in both cases Ca^{2+} binding produces a diffuse layer that is structurally and stereochemically matched.

However, if the Ca^{2+} concentration under the Langmuir monolayer is reduced to 5 mM, the bound Ca^{2+} ions are disordered and because the structural matching is no longer dominant, kinetic factors come into play and the less stable polymorph, vaterite, is nucleated. Interestingly, oriented vaterite crystals are also observed when the charge on the monolayer is reversed, for example by using a positively charged film of protonated octadecylamine ($CH_3(CH_2)_{17}NH_3^+$) molecules. In contrast, if only weak interactions exist at the monolayer–water interface, then no oriented nucleation occurs. For example, $CaCO_3$ crystallization is inhibited under neutral monolayers prepared from octadecanol ($CH_3(CH_2)_{17}OH$) or cholesterol ($C_{27}H_{45}OH$), and the crystals that form are no different from those in the control experiments.

The importance of electrostatic, structural and stereochemical interactions in the oriented nucleation of calcium carbonate crystals under compressed Langmuir monolayers is also borne out by studies of $BaSO_4$ crystallization under monolayers of long-chain alkyl sulfates, and ice or $CaSO_4 \cdot 2H_2O$ nucleation under monomolecular films of long-chain alcohols.

Langmuir monolayers induce the oriented nucleation of inorganic crystals through structural, electrostatic and stereochemical matching arising from ionic interactions with organized arrays of surfactant headgroups.

9.4 Morphosynthesis of biomimetic form

The morphogenesis of biominerals with complex, time-dependent form is dependent on the vectorial regulation of growth within shaped biological compartments, such as vesicles and macromolecular frameworks. We discussed several aspects of this process in Chapter 7 based around processes of physical, and to a lesser extent, chemical patterning. One major challenge in biomineral-inspired materials chemistry is the synthetic reproduction of analogous structures, using an approach that we shall refer to as *morphosynthesis*. In this section we describe some examples of how inorganic materials with complex morphologies can be produced. Two strategies, based on physical and chemical patterning, are illustrated (see Table 9.5).

The study of biomineral morphogenesis is inspiring the development of morphosynthetic routes to inorganic materials with complex form.

Table 9.5 Morphosynthesis of inorganic materials

Strategy	Approach	Product	System	Materials
Physical patterning	Supramolecular templates	Mineral tubes and helices	Lipid tubules	Cu, Ni, Al_2O_3, SiO_2
		Inorganic sponges	Copolymer gels	Fe_3O_4, TiO_2
			Bacterial threads	SiO_2, zeolites
	Reaction field replication	Hollow shells	Emulsion droplets	$CaCO_3$, SiO_2
		Cellular thin films	Microemulsion foams	$CaCO_3$, MnOOH, FeOOH
		Porous hollow shells	Foams + latex beads	$CaCO_3$, FeOOH
Chemical patterning	Reaction field instability	Micro-skeletal frameworks	Bicontinuous microemulsions	$Ca_{10}(OH)_2(PO_4)_6$, SiO_2
		Twisted/coiled filaments	Reverse microemulsions	$BaSO_4$, $BaCrO_4$, $CaSO_4$
		Nested filaments	Block copolymer micelles	Ca phosphates

9.4.1 Physical patterning with supramolecular templates

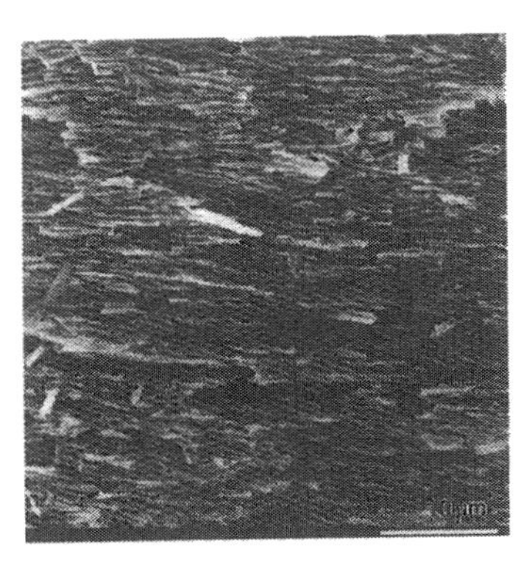

Fig. 9.23 Section cut parallel to the long axis of a silica-infiltrated bacterial thread after removal of the multicellular filaments, showing co-aligned channels. Scale bar, 10 μm.

Many biominerals are physically patterned inside vesicles held against scaffolds such as intracellular organelles, micro-skeletal filaments and cell wall membranes (see Chapter 7, Section 7.3.1). Analogous structures can be self-assembled in the test-tube and used as shaped templates for replication by inorganic mineralization. This approach therefore incorporates much of what we discussed in Section 9.3 with the additional feature that the organic template is usually removed by thermal degradation after mineralization to produce a pure inorganic replica. In general, shaped organic supramolecular objects, such as lipid tubules, are assembled prior to the formation of the inorganic phase, and nucleation and growth across and around the curved external surfaces produce an inorganic cast of the organic structure. Alternatively, organic structures with ordered internal spaces, such as polymer sponges or bacterial superstructures, can be used to produce inverse inorganic replicas of the internal architecture. For example, removal of the bacterial template from the fibre composites described in Section 9.2.3 by heating to 600°C produces an intact silica fibre that consists of thousands of 0.5-μm-wide channels co-aligned along the morphological long axis (Fig. 9.23).

In general, inorganic replicas with high fidelity are obtained only if the interactions at the surface of the organic template are competitive over analogous processes in bulk solution. This requires that the surface is relatively stable throughout the inorganic reaction and chemically functionalized for interfacial recognition. In addition, the template has to be removed without fracturing the cast and this is often difficult to achieve. Overall, therefore, *direct* physical patterning requires four principal steps:

template assembly → interfacial recognition →
mineral replication → template removal

One possible variation in this approach is to assemble the organic template *in situ* during inorganic precipitation such that the inorganic and organic components are patterned *synergistically*. This requires a high degree of structural and electrostatic complementarity at the inorganic–organic interface; otherwise the components phase separate and there is no structural coupling and associated mineral replication. Four steps are required to achieve high fidelity patterning:

complementarity → co-assembly → mineral replication →
template removal

Whereas direct patterning is based on a one-to-one correspondence between the organic template and inorganic replica—you get what you are given so to speak—a synergistic approach is more unusual in that the collective actions of the inorganic and organic components might produce something unexpected in terms of shape. For example, the phospholipid *diacetylenic phosphatidyl choline* ($DC_{8,9}PC$) has a zwitterionic headgroup and two hydrocarbon tails each containing a –C≡C–C≡C– unit between carbon atoms numbered 8 and 9, as shown in Fig. 9.24. Because the lipid has a chiral molecular structure, it self-assembles in water into a tubular structure.

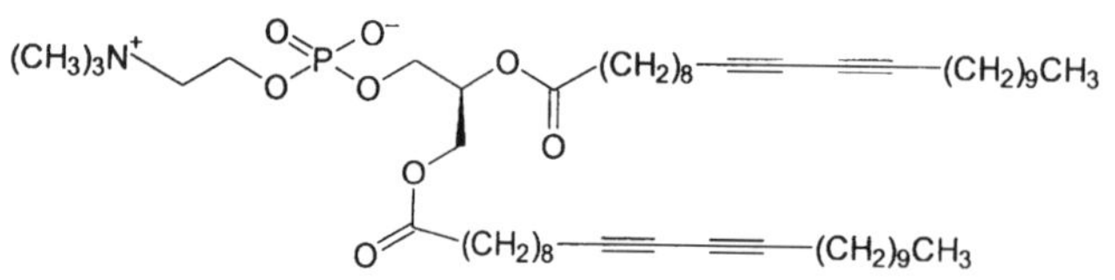

Fig. 9.24 Molecular structure of diacetylenic phosphatidyl choline ($DC_{8,9}PC$).

We described a similar behaviour for a galactocerebroside lipid in Section 9.3.2, but in the case of $DC_{8,9}PC$ the bilayers are initially formed as thin multilamellar strips rather than large sheets, and the stress field associated with packing of the chiral molecules is directed away from the long axis of each strip. This produces both turning and bending forces so that the strips are twisted into a helical ribbon, which subsequently transforms into a helical tube—a supramolecular drinking straw—as the edges of the helix become sealed together (Fig. 9.25).

Used in this form, the preformed tubules are excellent templates for the surface deposition of thin inorganic coatings—aluminium hydroxide, metallic nickel, colloidal silica for example—and subsequent preparation of hollow inorganic cylinders by burning out the lipid at elevated temperatures. In contrast, when diacetylenic phosphatidylcholine molecules are added to water containing tetraethoxysilane ($Si(OCH_2CH_3)_4$, TEOS), assembly of the helical tubes takes place at the same time that the TEOS molecules are hydrolysing to produce silicate anions. If the rates of self-assembly and hydrolysis are similar then silicate anions preferentially interact with the cationic $-N(CH_3)_3^+$ headgroups of the lipid molecules and they assemble together into a multilamellar helical ribbon. With time, the silicate anions begin to condense and a 3-nm-thick layer of silica is deposited between each of the twisted lipid bilayers. As shown in Fig. 9.26, the resulting silica–lipid nanocomposite often has an open helical form rather than the cylindrical architecture obtained by surface deposition of silica on preformed tubules.

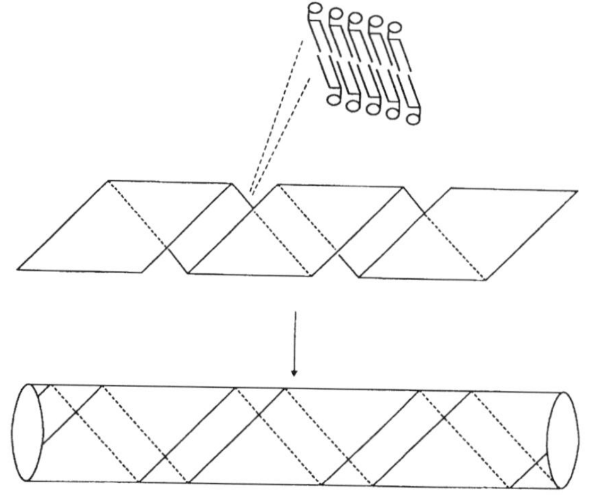

Fig. 9.25 Multilamellar helical ribbon and helical tube of $DC_{8,9}PC$.

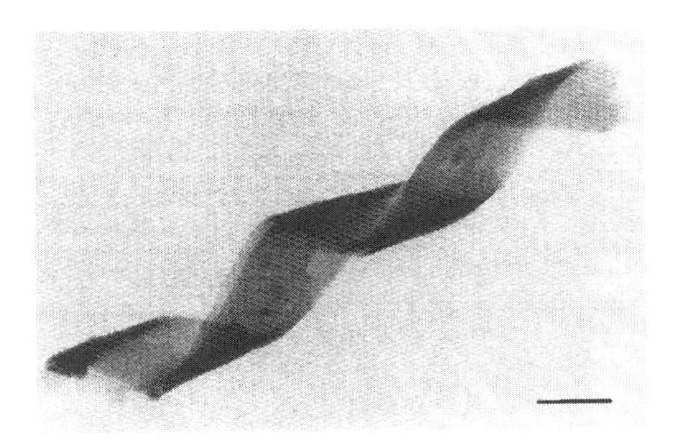

Fig. 9.26 Silica–lipid mineralized helical ribbon. Scale bar, 400 nm.

Supramolecular templates are used in the physical patterning of inorganic materials. In most cases preformed templates are used and the inorganic shape is produced by direct replication. Synergistic processes involving co-assembly of the template and inorganic components are also possible.

9.4.2 Physical patterning from reaction field replication

Many of the complex forms of biominerals derive from the shaping of boundary-organized reaction environments. For example, as we described in Chapter 7, Section 7.3, vesicles are shaped by microtubules that are organized radially and tangentially in the cell. The mineral is then physically patterned within the vesicle as a consequence of the restrictions placed on the directions of growth in the shaped reaction field. Alternatively, the reaction environment can be organized as the interstitial spaces within an assembly of vesicles or cells. In Chapter 7, Section 7.3.2 for example, we described how vesicle foams are responsible in part for the patterning of the elaborate porous silica frameworks of diatoms and radiolarians. In this section, we illustrate how synthetic counterparts of these organized or shaped reaction fields can be used to produce inorganic materials with biomimetic form (see Table 9.5).

We start with the sphere as the simplest shape adopted by a boundary-organized reaction field. Droplets of oil can be suspended in water by adding a suitable surfactant that stabilizes the oil–water interface (Fig. 9.27A). This is basically the process involved in making vinaigrette. Such systems are termed

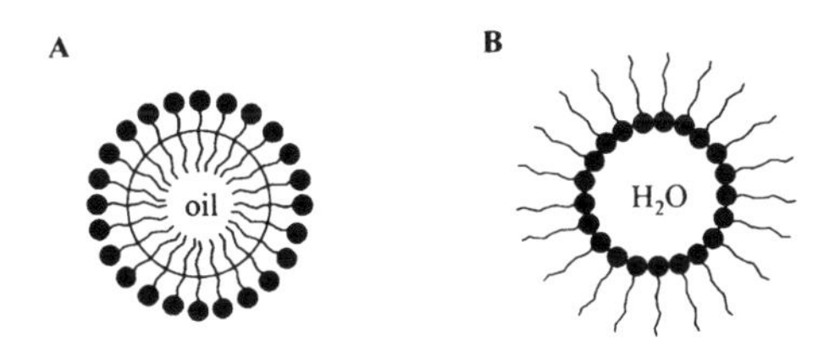

Fig. 9.27 (A) Oil-in-water microemulsions; (B) water-in-oil reverse microemulsions.

microemulsions and the maximum size of the droplets depends on how much oil can be stabilized by the surfactant before the drops begin to coalesce and separate out as an immiscible layer. A mineralized replica of the outer surface of the drop is produced when the nucleation and growth of an inorganic material occur preferentially at the oil–water interface. For example, if the oil capsules contain molecules such as tetraethoxysilane then the slow hydrolysis of the organosilane at the oil–water interface results in the formation of a hollow sphere of silica around the surface of each oil droplet. By changing the size of the droplets, shells of different diameter can be prepared.

A similar approach can also be used with *reverse microemulsions*, which consist of surfactant-stabilized water droplets suspended in oil (Fig. 9.27B). For example, microemulsions containing hexane, sodium dodecylsulfate ($CH_3(CH_2)_{11}OSO_3H$) and aqueous droplets of calcium bicarbonate prepared as described in Section 9.3.3 become supersaturated when CO_2 is removed from the water phase. Although CO_2 cannot escape easily to the atmosphere because of the surrounding oil, the water droplets become partially degassed because microbubbles of CO_2 nucleate at the hydrophobic boundary around each droplet. This increases the local supersaturation with the consequence that crystals of vaterite ($CaCO_3$) are deposited around the gas bubbles trapped at the oil–water interface (Fig. 9.28). With time, the accumulation of $CaCO_3$ results in a hollow inorganic shell with unusual surface pores and indentations (Fig. 9.29).

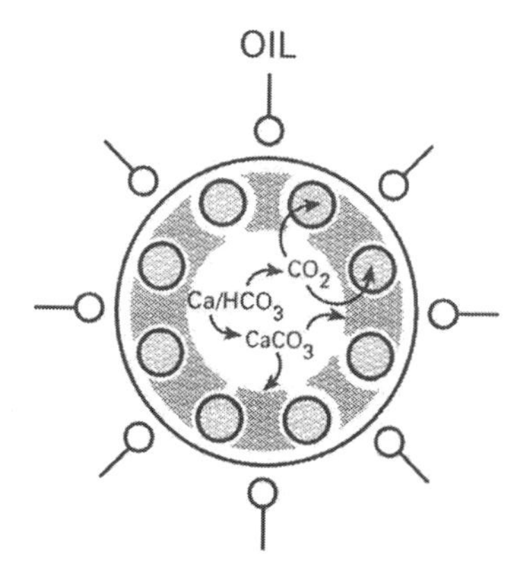

Fig. 9.28 Growth and patterning of calcium carbonate hollow shells in reverse microemulsions.

Under certain conditions, increasing the concentration of oil droplets in a microemulsion produces a *biliquid foam* in which closely packed oil droplets are stabilized by a thin soapy aqueous film and suspended in a continuous phase of water. If the water phase is replaced by a supersaturated solution such that an inorganic mineral grows in the interstitial spaces of the foam then the system loosely resembles the patterning process we described in Chapter 7, Section 7.3.2, for the silica shells of diatoms and radiolarians. From a materials perspective, turning the reaction field of the foam into stone would provide a simple direct route to inorganic solids with cellular or honeycombed architectures.

One way to produce a biliquid foam is to spread a thin film of a supersaturated microemulsion on a flat metal substrate and partially remove the oil

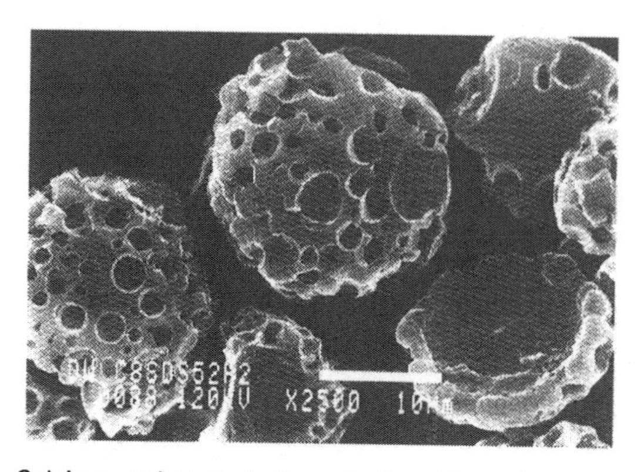

Fig. 9.29 Calcium carbonate hollow shells with surface pores. Note also the presence of a broken shell. Scale bar, 10 μm.

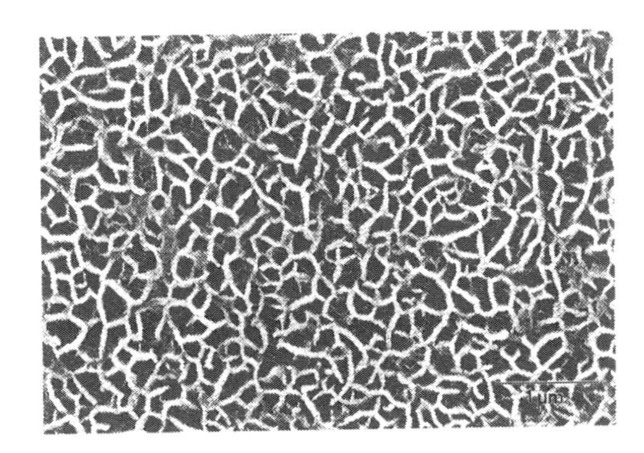

Fig. 9.30 Cellular film of calcium carbonate prepared in a biliquid foam. Scale bar, 1 μm.

phase by washing with hot hexane. This destabilizes the microemulsion film and causes the remaining components to separate into a foam-like array of tiny oil droplets surrounded by the supersaturated aqueous fluid. Growth of inorganic crystals then occurs in the interstitial spaces and boundary edges between the oil droplets to produce a mineralized thin film imprinted with a cellular structure. This approach has been used to prepare disordered frameworks of calcium carbonate (aragonite) and transition metal oxides (FeOOH, MnOOH), as shown in Fig. 9.30. Typically, the cellular films have continuous and branched mineral walls, 20 to 100 nm in width, and cell sizes between 45 and 300 nm, depending on the size of the oil droplets. Because the foam is a transitory structure, mineralization and oil droplet self-assembly must occur almost simultaneously if the interstitial spaces are to be filled with a continuous inorganic framework of calcium carbonate or Fe^{III} oxide. This is achieved, respectively, by CO_2 outgassing of a calcium bicarbonate solution or O_2 diffusion into an Fe^{II}-containing microemulsion, both of which only occur when the area of the air–water interface increases significantly during foam formation in the thin film.

If the flat metal substrate used in the above procedure is replaced with a polymer bead then the resulting mineralized cellular films can be sculpted into three dimensions. For example, spreading a microemulsion film containing supersaturated calcium carbonate over micrometre-sized polystyrene beads followed by washing in hot hexane induces foam formation and concomitant mineralization of a cellular structure around the polymer particles. The beads are then dissolved in chloroform or destroyed by heating to give porous hollow shells of cellular calcium carbonate that look like artificial versions of biomineralized micro-skeletons (Fig. 9.31).

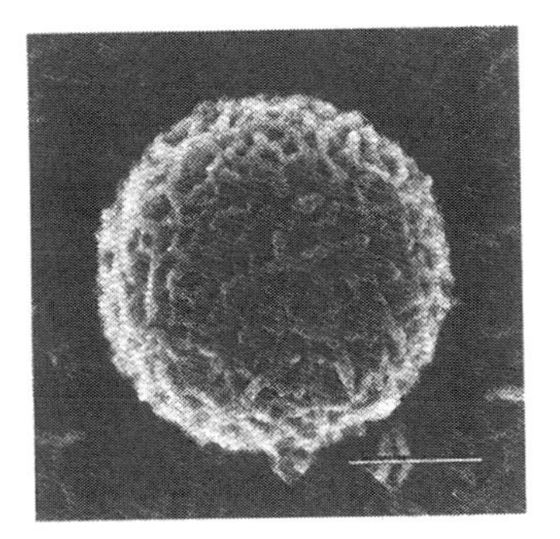

Fig. 9.31 Calcium carbonate porous hollow shell with cellular surface structure. Scale bar, 500 nm.

Inorganic morphosynthesis of porous shells and membranes involves the materials replication of stable or transitory reaction fields established in microemulsion droplets and biliquid foams.

9.4.3 Chemical patterning in unstable reaction fields

In biomineralization, chemical patterning generally refers to a vectorial process of morphogenesis in which supersaturation levels are regulated in time and space by changes in the activity and positioning of ion pumps, channels or cells responsible for the supply of ions to the mineralization front (see Chapter 7, Section 7.2.1). Although it is not easy to translate this process into an analogous counterpart in synthetic materials chemistry, the dynamical aspects of chemical patterning can be reproduced in the laboratory to prepare some amazing structures with complex morphological form. To achieve this, the physical replication of a confined reaction field is coupled with *instability thresholds* in the shape of the surrounding environment brought about by the onset of mineralization. Under these conditions, the direct correspondence between the original shape and size of the fluid-filled compartments and that of the final mineral phase, as described in Section 9.4.2, breaks down because development of the inorganic structure perturbs the local environment in which it grows. This can result in symmetry breaking (Chapter 7, Section 7.1) because as the reaction field adjusts to the presence of the incipient mineral then this in turn influences new growth directions and a feedback loop is established. The inorganic morphology therefore becomes dependent on the interplay of these processes and how they emerge with time and length scale.

Instability thresholds can be breached in surfactant and polymer micelles and microemulsions by *in situ* inorganic precipitation to produce materials with time-dependent morphologies (Table 9.5). In some cases, the mineral morphology superficially resembles the reaction field but differs dramatically when compared in terms of scale, whereas in other systems, there appears to be no correspondence at all. An example of each is described below.

Although microemulsions are usually in the form of spherical droplets of water in oil, or oil in water, some surfactants can stabilize the separation of the oil and water components into two interpenetrating networks of highly branched continuous nanoscopic conduits (Fig. 9.32). These compartmental-

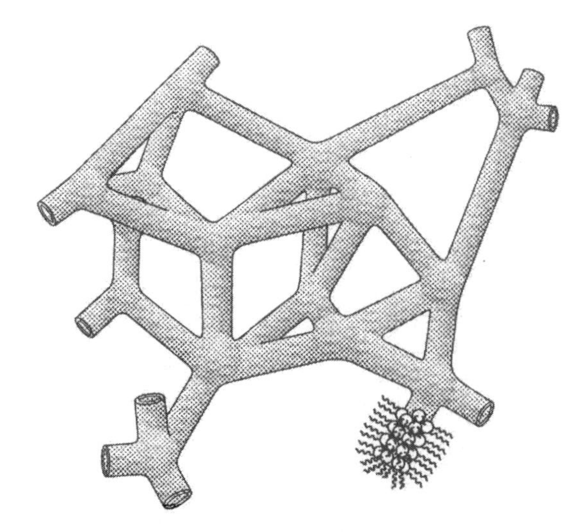

Fig. 9.32 Idealized structure of a bicontinuous microemulsion. For clarity, only the nanoscopic water channels are shown along with surfactant molecules. The oil channels are of similar dimension and interpenetrate the water conduits.

ized liquids are referred to as *bicontinuous microemulsions*, and can be prepared, for example, from mixtures of tetradecane ($C_{14}H_{30}$), water and the cationic surfactant didodecyldimethylammonium bromide (DDAB, $[CH_3(CH_2)_{11}]_2N(CH_3)_2{}^+Br^-$). By using a supersaturated calcium phosphate solution in place of pure water, and freezing the tetradecane channels at temperatures above 0°C—the oil channels are highly mobile at room temperature—we might expect the plumber's nightmare of water-filled pipes to transcribe into a nanoporous inorganic replica. But this only occurs at the beginning of calcium phosphate precipitation. Instead, the incipient crystals perturb the reaction field by influencing the local structure of the microemulsion, for example through adsorption of water and surfactant molecules onto the mineral surface. This results in localized changes in the scale of the water compartments such that a remarkable calcium phosphate micro-skeletal architecture is formed after several days (Fig. 9.33). The structure consists of an interconnected framework of hydroxyapatite needle-shaped crystals, and does therefore bear some resemblance to the structure of the reaction field associated with the bicontinuous microemulsion. But the crystals are approximately 100 nm in thickness, which is about fifty times wider than the aqueous channels of the reaction environment. So although the inorganic structure is sculpted in part by reactions in the confined nano-channels, chemical interactions between the mineral and microemulsion during crystal growth impose structural changes that are ultimately expressed on a longer length scale.

In contrast, the growth of $BaSO_4$ crystals in water-in-oil spherical microemulsions prepared from the anionic surfactant sodium bis(2-ethylhexyl)sulfosuccinate (commonly called AOT, see Fig. 9.34) produces structures that bear no resemblance to the shape of the reaction environment. In this case, the reaction occurs in spherical water droplets only 4 nm across but produces micrometre-long twisted bundles of $BaSO_4$, as shown in Fig. 9.35. The reaction occurs at room temperature in unstirred isooctane containing a mixture of two different reverse microemulsions—one consisting of hydrated $Ba(AOT)_2$ and another prepared from NaAOT with encapsulated sulfate anions. When mixed together, the microemulsions slowly exchange their con-

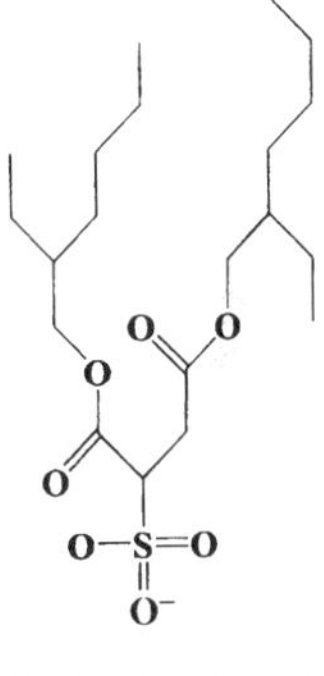

Fig. 9.34 Molecular structure of sodium bis(2-ethylhexyl)sulfosuccinate (AOT).

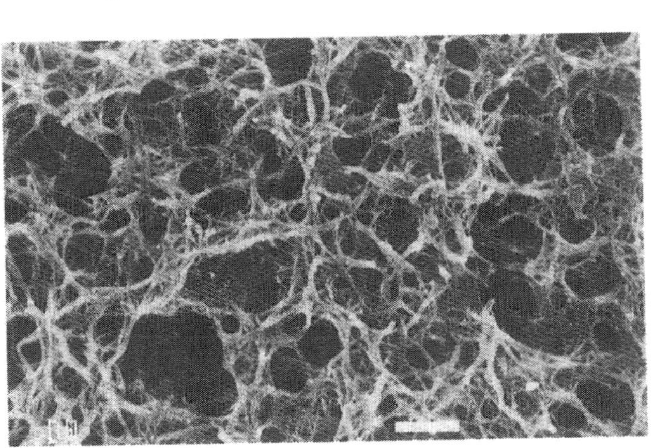

Fig. 9.33 Micro-skeletal form of calcium phosphate prepared in bicontinuous microemulsions. Scale bar, 1 μm.

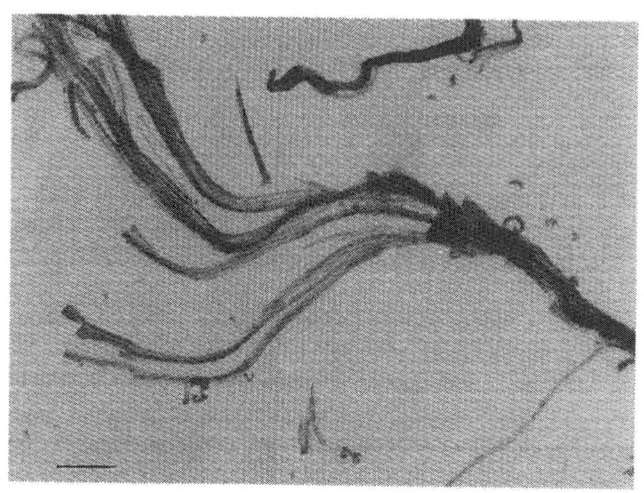

Fig. 9.35 Twisted bundles of $BaSO_4$ nanofilaments prepared in AOT reverse microemulsions. Scale bar, 500 nm.

tents and $BaSO_4$ precipitates within the water-filled pools. If there is a stoichiometric excess of Ba^{2+} ions then the initially formed nanoparticles are positively charged and the surfactant molecules are strongly adsorbed onto the inorganic surfaces. This has the effect of producing amorphous rather than crystalline nanoparticles of $BaSO_4$ that are limited to 5 nm in size and slowly aggregate because of hydrophobic interactions between the surface-anchored surfactant chains. Amazingly, groups of nanoparticles line up in linear arrays that fuse together and crystallize into 5-nm-wide single $BaSO_4$ filaments (steps A and B in Fig. 9.36). With time, other crystalline filaments are formed parallel to the original thread to produce a small bundle held together by surfactant bilayers (Fig. 9.36, step C). The locking in of new filaments by surfactant interdigitation generates a bending force in the bundle to give a coiled

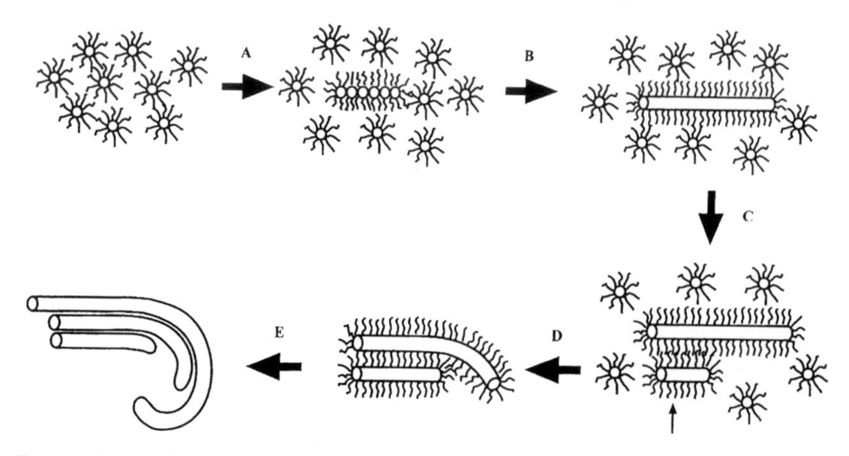

Fig. 9.36 Mechanism for the formation of twisted bundles of $BaSO_4$ nanofilaments from surfactant-stabilized amorphous nanoparticles. Small arrow shows nucleation of a secondary filament and surfactant interdigitation. See text for details.

spiral at one end (steps D and E in Fig. 9.36). Sometimes this force is perpetuated throughout the whole length of the bundle and $BaSO_4$ structures with helical form are produced. As the bundles thicken, packing pressures force the filaments to splay outwards and cone-shaped structures are produced at the growth tips as shown in Fig. 9.35.

The morphosynthesis of hydroxyapatite and $BaSO_4$ crystals with micrometre-scale complex forms from nanometre-sized reaction fields is surprising and requires a lot more research to clarify the mechanisms. More details, as well as information on the use of polymer-based micelles in calcium phosphate morphosynthesis, can be found in the suggestions for further reading.

Mineral-induced instabilities in synthetic reaction fields, such as channel-like or spherical microemulsions, give rise to the emergence of complex inorganic morphologies that change with time and length scale.

9.5 Crystal tectonics

The amazing ability of organisms to construct higher-order biomineralized structures from preformed mineral-based building blocks was discussed in detail in Chapter 8. In general, biomineral tectonics involves multiple levels of processing (see Chapter 8, Section 8.4), in which inorganic crystals and amorphous particles are prefabricated in vesicles (Section 8.2) and then translocated to remote mineralization sites for higher-order assembly (Section 8.3). An analogous approach—*crystal tectonics*—attempts to couple the synthesis and self-assembly of inorganic nanoparticle building blocks to give artificial structures with controlled organization. For this, the surfaces of the synthesized particles have to be marked with chemical information that is addressable under certain conditions of higher-order assembly. Here we describe two approaches. First, we discuss how organized arrays can be directly produced in reaction solutions by *interactive assembly* involving the interdigitation of surfactant molecules that become attached to nanoparticle surfaces during inorganic synthesis. The driving force for assembly is the formation of a bilayer between adjacent particles and this becomes directional if the organic molecules are located on specific crystal faces. Because the coupling of synthesis and self-assembly is often difficult to achieve in practice, an alternative approach is to attach complementary ligands to the surfaces of preformed nanoparticles. This has the advantage that the informational content is *programmed* into the surfaces of the inorganic building blocks for recognition-driven assembly. For example, the highly specific recognition properties of antibodies and antigens, proteins and ligand substrates or between complementary single strands of DNA are being used in crystal tectonics.

Crystal tectonics can be defined as the chemical construction of higher-order structures from solid-state building blocks, such as inorganic nanoparticles. Complementary interactions between chemical groups on the surfaces of the individual nanoparticles are responsible for ordered aggregation.

9.5.1 Interactive assembly

At first sight it seems difficult to reconcile the complex and dynamical processes of biomineral tectonics with the static methods usually employed in inorganic materials synthesis. But there are some promising signs that this might be possible. For example, in Section 9.4.3 we described how surfactant-capped $BaSO_4$ clusters spontaneously ordered into linear aggregates and fused together to produce thin crystalline filaments. If we can prevent the fusion into a continuous structure, then in principle we would have a linear array of membrane-bounded nanoparticles similar to what is observed, for example, in the magnetite chains of magnetotactic bacteria described in Chapter 2 (see Section 2.6.1 and Fig. 2.21).

The trick is to increase the size and stability of the nanoparticles present in the synthesis mixture before aggregation occurs. It turns out that discrete crystalline $BaSO_4$ nanoparticles are only formed in the reverse microemulsions described in Section 9.4.3 if the anion concentration is two to five times that of the Ba^{2+} cations. This is because under these conditions the charge on the $BaSO_4$ crystals is negative due to a surface excess of anions, and there is therefore minimal interaction with the anionic headgroups of the bis (2-ethylhexyl) sulfosuccinate (AOT) surfactant molecules. The surfactant molecules therefore remain mobile and no particle aggregation is induced. On the other hand, as we described in Section 9.4.3, an excess of Ba^{2+} gives rise to unstable amorphous nanoparticles that fuse into crystalline nanofilaments because the surfactant molecules are anchored strongly to the inorganic surfaces by electrostatic forces (see Fig. 9.36). So, a compromise between these two sets of conditions should be capable of producing larger nanoparticles with increased stability, but which slowly associate into a higher-order structure because some, but not all, of the surfactant molecules are anchored to the inorganic surface.

Indeed, this is exactly what happens when the Ba^{2+} to SO_4^{2-} molar ratio is set at 1:1. Under these conditions, *linear chains* of individual $BaSO_4$ nanoparticles are formed spontaneously in the reverse microemulsions within a few hours of mixing the reactants, and identical structures consisting of $BaCrO_4$ nanoparticles are produced when microemulsions of $Ba(AOT)_2$ and chromate-containing NaAOT are mixed (Fig. 9.37). In both cases, the colloidal chains are 50 to 500 nm in length, and each chain consists of discrete rectangular prismatic crystals that are uniform in size (*ca.* $16 \times 6.8 \times 6$ nm) and preferentially aligned with the long axis of each particle perpendicular to the chain direction. A regular spacing of 2 nm, corresponding to an interdigitated surfactant bilayer, separates each crystal in the chain so that they look like tiny backbones when viewed in the electron microscope.

The linear array is dependent on the uniformity in nanoparticle size and the presence of crystal faces with regular shape. These facilitate crystal face-specific interactions between the hydrophobic tails of AOT molecules adsorbed onto the flat side faces of the prismatic nanoparticles so that a bilayer of interdigitated surfactant molecules connects the particles in the chain (Fig. 9.38). This process occurs specifically along one crystallographic direction because there are two sets of side faces that differ slightly in surface area. In fact, the aggregation process is directed along an axis normal to the largest side face because this maximizes the hydrophobic–hydrophobic interactions between the crystals and lowers the free energy of the self-assembled structure.

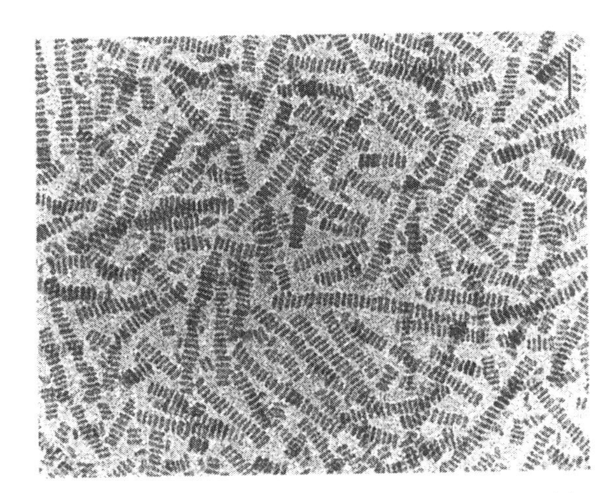

Fig. 9.37 Linear chains of periodically arranged $BaCrO_4$ nanoparticles prepared in AOT reverse microemulsions. Scale bar, 50 nm.

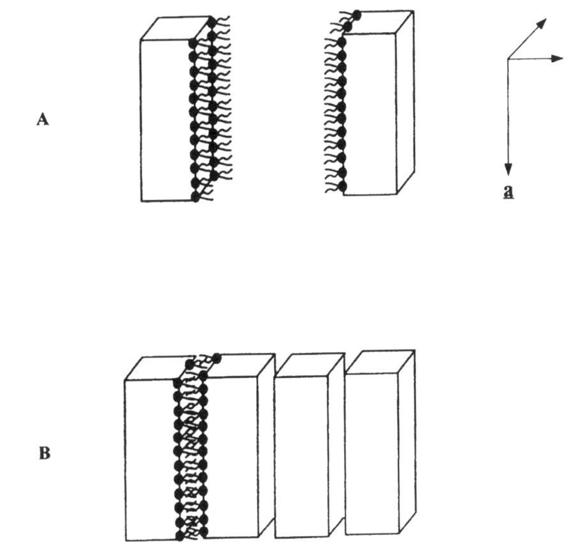

Fig. 9.38 Mechanism for the face-specific linear aggregation of $BaCrO_4$ or $BaSO_4$ nanoparticles. (A) Individual particles with surface-anchored AOT molecules. Only one face is shown covered with the surfactant molecules for clarity. The direction of the crystallographic *a* axis is shown. (B) Interdigitation of the surfactant chains and chain formation.

The interfacial activity of microemulsion-based reaction fields is used to interactively couple nanoparticle synthesis and self-assembly to produce higher-order structures with linear chain architecture.

9.5.2 Programmed assembly

The highly specific recognition properties of *antibodies and antigens* make them excellent candidate molecules for the programmed assembly of preformed nanoparticles in solution. Antibody engineering is a highly selective method for making proteins with binding sites for specific molecules. For

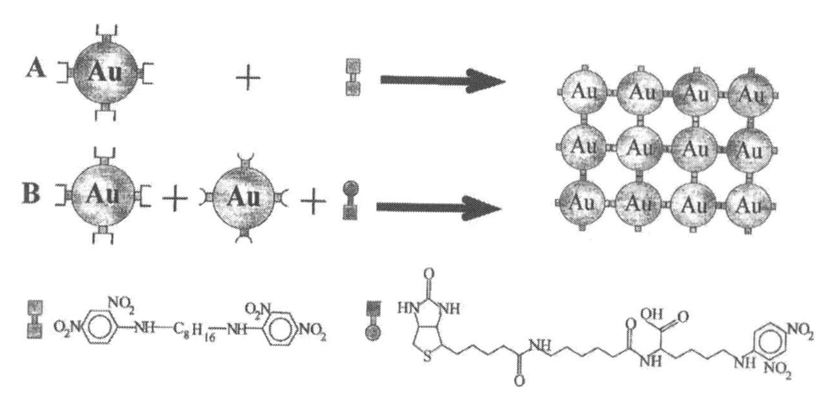

Fig. 9.39 Antibody–antigen coupling of Au nanoparticles: (A) surface-adsorbed anti-DNP antigen and DNP–DNP antigen; (B) surface-adsorbed anti-DNP or anti-biotin antibodies, and biotin–DNP antigen. Molecular structures of the antigens are also shown.

example, if an antibody such as *immunoglobulin E* (IgE) is raised against a molecule containing a dinitrophenyl (DNP) group—the resulting antibody is referred to as anti-DNP IgE—then the protein will contain clusters of amino acid side chains that specifically recognize and bind DNP groups associated with molecules in solution. The high binding constant associated with this process can be exploited in crystal tectonics by adsorbing the anti-DNP antibody onto Au nanoparticles and then using a specially designed ligand molecule with a double-headed DNP functionality—the antigen—to reversibly aggregate the particles. This is shown in Fig. 9.39A. Interparticle cross-linking only occurs if the DNP headgroups are separated by a spacer chain of at least eight carbon atoms so that antibodies on adjacent nanoparticles can be coupled together. The aggregation process then becomes highly specific and a nanoparticle-containing precipitate is obtained (Fig. 9.40). Similar experiments have been undertaken using surface-adsorbed antibodies specifically engineered for the binding of *biotin* headgroups. Moreover, as shown in Fig. 9.39B, mixtures of different Au nanoparticles with either surface-adsorbed anti-DNP or anti-biotin immunoglobulins can be reversibly aggregated if a synthetic antigen with DNP and biotin groups at alternate ends of the molecule is added to the colloidal sol. As before, a spacer group is required—in this case 19 atoms long—so that both types of antibody can be accessed by the cross-linking ligand.

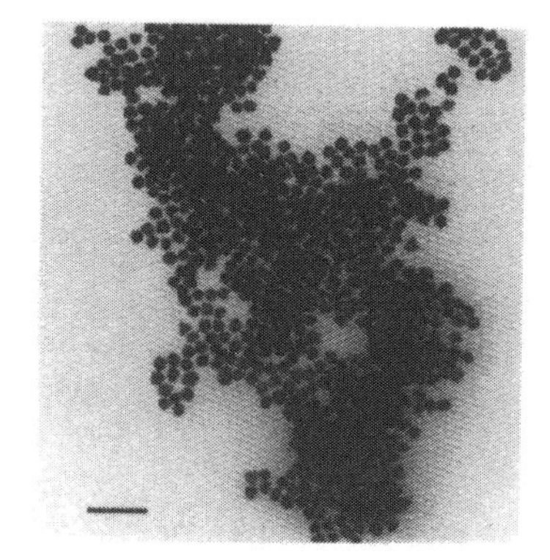

Fig. 9.40 Antibody–antigen coupled Au nanoparticle assemblies. Scale bar, 60 nm.

Biotin also binds very strongly to the protein *streptavidin*, and this has also been used to assembly Au nanoparticles. Reversible cross-linking of biotinylated groups adsorbed onto the nanoparticle surface occurs because there are four binding sites in each streptavidin molecule so there are enough connecting units for 3-D aggregation. The surface attachment of biotin ligands is often difficult and one approach around this problem is to use nanoparticles encapsulated in a protein or lipid shell that can be covalently functionalized with biotin surface groups. For example, 60 to 70 exposed lysine residues on the surface of the protein shell of *ferritin* (see Section 9.2.2) can be readily coupled with long-chain biotinylated compounds, such

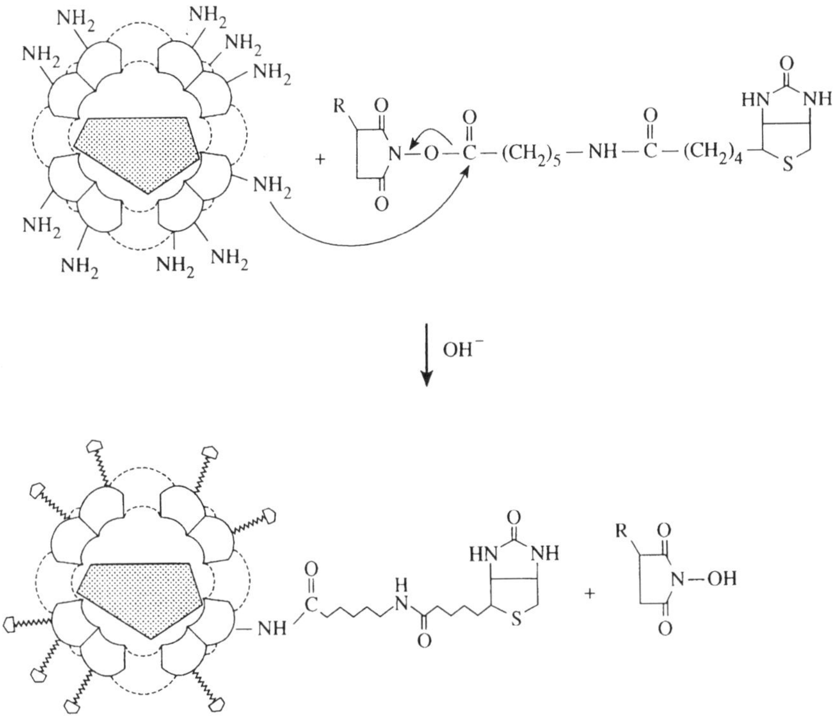

Fig. 9.41 Synthesis of biotinylated ferritin.

as shown in Fig. 9.41. The resulting biotinylated ferritin is conjugated in solution with streptavidin to produce an interconnecting network of iron oxide nanoparticles (Fig. 9.42). A flexible hexanoate spacer is incorporated into the biotin linkage to access the binding sites in streptavidin. This works well if there is a balance between the number of free and bound biotin groups on the protein surface. If too much streptavidin is added then all the binding sites are saturated before cross-linking of the protein shells can occur.

Finally, single-stranded oligonucleotides with complementary nucleotide bases can be attached to different populations of Au nanoparticles. When these are mixed, those particles with complementary strands are specifically aggregated due to base pairing and duplex formation between the particles. The method is highly selective in controlling the distances between the aggregated nanoparticles because these depend on the number of bases in the synthesized oligonucleotide. Moreover, it can be used for the selective extraction of nanoparticles from heterogeneous populations as every particle is coded with its own brand of DNA.

At the current time, however, no ordered superstructures, analogous to the linear chains described in Section 9.5.1, have been produced by programmed assembly. This is because spherical nanoparticles have been used. In the future, it should be possible to use nanoparticles with geometric shapes and regular crystal surfaces for directional coupling. Interesting superstructures are expected, particularly if the complementary recognition pairs are attached to specific crystal faces.

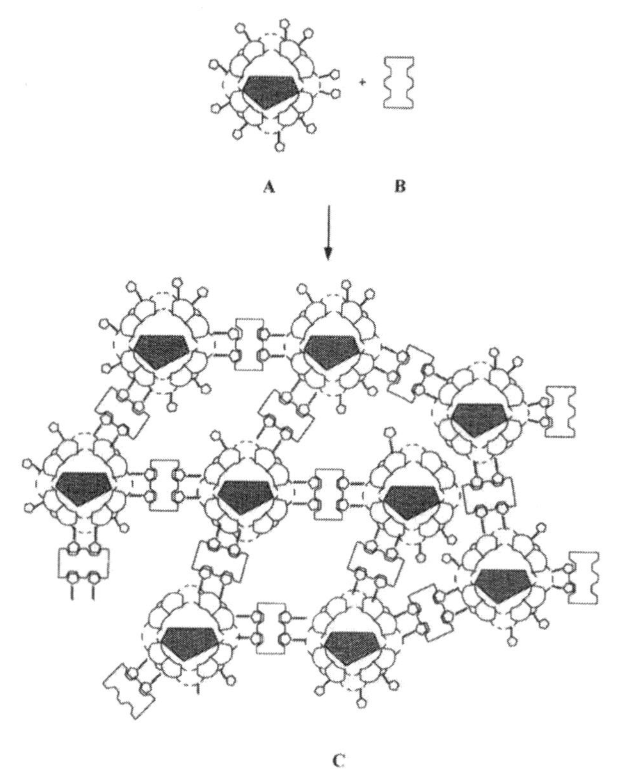

Fig. 9.42 Coupling of biotinylated ferritin (A) with streptavidin (B) to produce a cross-linked network of protein-coated iron oxide nanoparticles (C).

Ligand-induced recognition and binding of antibody–antigen, protein–substrate or oligonucleotide complementary pairs induce the reversible aggregation of inorganic nanoparticles.

9.6 Summary

In the final chapter of this book we have discussed how biomineralization processes can be related to new synthetic strategies in materials chemistry. The focus has been on illustrating how four key biomineralization concepts that we discussed in previous chapters—supramolecular preorganization, interfacial molecular recognition, vectorial regulation and multilevel processing—are inspiring biomimetic approaches. These included, respectively, synthesis in confined reaction spaces, template-directed materials synthesis, morphosynthesis of biomimetic form, and construction of higher-order structures by crystal tectonics.

Much progress has been made in using supramolecular structures, such as unilamellar vesicles and artificial ferritins, to restrict the reaction spaces available for nanoparticle synthesis. The product is usually in the form of discrete inorganic particles that are encapsulated in a lipid or protein membrane. On the other hand, more complex materials and composites are being produced by using confined reaction spaces that are organized inside extended structures, such as multilamellar vesicles, bacterial threads and polymer sponges.

Supramolecular structures can also be used as templates for the controlled nucleation and growth of inorganic materials, provided that the organic surfaces contain functional groups that can interact with ions present in supersaturated solutions. For example, we discussed how the external surfaces of supramolecular lipid tubules are potent nucleators of iron oxides if sulfated lipids are included in the structure. Although the inorganic crystals are not crystallographically organized, other systems are available to study the underlying mechanisms that lead to oriented nucleation. In particular, compressed monolayers of simple surfactants induce the oriented nucleation of many different types of inorganic crystals if there is a high level of structural, electrostatic or stereochemical recognition at the template–mineral interface.

When the organic templates are shaped—in the form of lipid tubules for example—they can be exploited for the physical patterning of inorganic materials. Usually the organic structures are self-assembled and then added to supersaturated solutions, although they can also be assembled *in situ* during inorganic precipitation. Another promising approach in morphosynthesis involves the materials replication of stable or transitory reaction fields, such as microemulsion droplets and biliquid foams. When relatively stable, these give rise to porous shells and cellular membranes, respectively, but when destabilized by the mineralization reactions, complex forms with little resemblance to the shape and size of the reaction field are produced. For example, we discussed the formation of micrometre-scale frameworks of calcium phosphate in bicontinuous microemulsions that consisted of nanometre-sized water and oil channels, and micrometre-long twisted bundles of $BaSO_4$ nanofilaments in reverse microemulsions comprising spherical water droplets only 4 nm in diameter.

In the last section of the chapter we introduced two strategies in crystal tectonics that might begin to go some way towards synthetically constructing inorganic materials with higher levels of order. We discussed how nanoparticles are assembled spontaneously by surface-induced interactions that arise during synthesis or programming of preformed particles. In the former, hydrophobic forces between surfactant chains anchored to prismatic $BaSO_4$ or $BaCrO_4$ crystallites during their synthesis in water-in-oil microemulsions were sufficient to direct the formation of linear chains of periodically spaced nanoparticles. In contrast, the second approach was dependent on ligand-induced recognition between surface-adsorbed antibodies and designer antigens in solution, or streptavidin and surface-anchored biotin, or alternatively, DNA-based interactions using complementary strands of oligonucleotides bound to Au nanoparticles.

Although the examples described in this chapter are just a beginning in the development of biomineral-inspired materials chemistry, there can be no doubt that the study of the inorganic structures of life will continue to extend our vision and imagination of what is possible in the laboratory. Nature teaches us that the material world can be transmuted, that the seemingly inert and immutable properties of inorganic minerals can be instilled with life.

Further reading

Addadi, L. and Weiner, S. (1985). Interactions between acidic proteins and crystals: stereochemical requirements in biomineralization. *Proc. Natl. Acad. Sci. U.S.A.*, **82**, 4110–4114.

Addadi, L., Moradian, J., Shay, E., Maroudas, N. G. and Weiner, S. (1987). A chemical model for the cooperation of sulfates and carboxylates in calcite crystal nucleation: relevance to biomineralization. *Proc. Natl. Acad. Sci. U.S.A.*, **84**, 2732–2736.

Aksay, I. A., Trau, M., Manne, S., Honma, I., Yao, N., Zhou, L. *et al.* (1996). Biomimetic pathways for assembling inorganic thin films. *Science*, **273**, 892–898.

Antonietti, M., Breulmann, M., Goeltner, C. G., Coelfen, H., Wong, K. K. W., Walsh, D. *et al.* (1998). Inorganic–organic mesostructures with complex morphologies: precipitation of calcium phosphate in the presence of double-hydrophilic block copolymers. *Chem. Eur. J.*, **4**, 2491–2498.

Birchall, J. D. (1989). The importance of the study of biominerals to materials technology. In *Biomineralization: chemical and biochemical perspectives* (ed. Mann, S., Webb, J. and Williams, R. J. P.), pp. 491–509. VCH Verlagsgesellschaft, Weinheim.

Breulmann, M., Davis, S. A., Mann, S., Hentze, H. P. and Antonietti, M. (2000). Polymer gel templating of porous inorganic macrostructures using nanoparticle building blocks. *Adv. Mater.*, **12**, 502–507.

Calvert, P. (1995). Biomimetic ceramics and hard composites. In *Biomimetics: design and processing of materials* (ed. Sarikaya, M. and Aksay, I. A), pp. 145–161. American Institute of Physics series in polymers and complex materials. AIP Press, Woodbury, NY.

Douglas, D. (1996). Biomimetic approaches to nanoscale fabrication. In *Biomimetic materials chemistry* (ed. Mann, S.), pp. 117–142. VCH, New York.

Douglas, T. (1996). Biomimetic synthesis of nanoscale particles in organized protein cages. In *Biomimetic materials chemistry* (ed. Mann, S.), pp. 91–115 VCH, New York.

Douglas, T. and Young, M. (1998). Host–guest encapsulation of materials by assembled virus protein cages. *Nature*, **393**, 152–155.

Falini, G., Albeck, S., Weiner, S. and Addadi, L. (1996). Control of aragonite or calcite polymorphism by mollusc shell macromolecules. *Science*, **271**, 67–69.

Firouzi, A., Kumar, D., Bull, L. M., Besier, T., Sieger, S., Huo, Q. *et al.* (1995). Cooperative organization of inorganic-surfactant and biomimetic assemblies. *Science*, **267**, 1138–1143.

Heuer, A. H., Fink, D. J., Laraia, V. J., Arias, J. L., Calvert, P. D., Kendall, K. *et al.* (1992). Innovative materials processing strategies: a biomimetic approach. *Science*, **225**, 1098–1105.

Heywood, B. R. and Mann, S. (1994). Template-directed nucleation and growth of inorganic materials. *Adv. Mater.*, **6**, 9–20.

Hirai, T., Hariguchi, S., Komasawa, I. and Davey, R. J. (1997). Biomimetic synthesis of calcium carbonate particles in a pseudovesicular double emulsion. *Langmuir*, **13**, 6650–6653.

Kniep, R. and Busch, S. (1996). Biomimetic growth and self-assembly of fluoroapatite aggregates by diffusion into denatured collagen matrices. *Angew. Chem., Int. Ed. Engl.*, **35**, 2624–2626.

Kuhn, L., Fink, D. J. and Heuer, A. H. (1996). Biomimetic strategies and materials processing. In *Biomimetic materials chemistry* (ed. Mann, S.), pp. 41–68. VCH, New York.

Mann, S. (1993). Molecular tectonics in biomineralization and biomimetic materials chemistry. *Nature*, **365**, 499–505.

Mann, S. (1996). *Biomimetic materials chemistry*. VCH, New York.

Mann, S. (2000). The chemistry of form. *Angew. Chem., Int. Ed. Engl.*, **39**, 3392–3406.

Mann, S. and Ozin, G. A. (1996). Synthesis of inorganic materials with complex form. *Nature*, **382**, 313–318.

Mann, S., Hannington, J. P. and Williams, R. J. P. (1986). Phospholipid vesicles as a model system for biomineralization. *Nature*, **324**, 565–567.

Mann, S., Burkett, S. L., Davis, S. A., Fowler, C. E., Mendelson, N. H., Sims, S. D. *et al.* (1997). Sol–gel synthesis of organized matter. *Chem. Mater.*, **9**, 2300–2310.

Mann, S., Davis, S. A., Hall, S. R., Li, M., Rhodes, K. H., Shenton, W. *et al.* (2000). Crystal tectonics: chemical construction and self-organization beyond the unit cell. *Dalton Trans.*, 3753–3763.

Ogasawara, W., Shenton, W., Davis, S. A. and Mann. S. (2000). Template mineralization of ordered macroporous chitin–silica composites using cuttlebone-derived organic matrix. *Chem. Mater.*, **12**, 2835–2837.

Ozin, G. A. (1992). Nanochemistry: synthesis in diminishing dimensions. *Adv. Mater.*, **4**, 612–649.

Ozin, G. A. (1997). Morphogenesis of biomineral and morphosynthesis of biomimetic forms. *Acc. Chem. Res.*, **30**, 17–27.

Sarikaya, M. and Aksay, I. A. (ed.) (1995). *Biomimetics: design and processing of materials*. American Institute of Physics series in polymers and complex materials. AIP Press, Woodbury, NY.

Tanev, P. T. and Pinnavaia, T. J. (1996). Biomimetic templating of porous lamellar silicas by vesicular surfactant assemblies. *Science*, **271**, 1267–1269.

Index

18395684R00121

Made in the USA
Middletown, DE
10 March 2015